Project California:
a Data Center Virtualization Server

Unified Computing System (UCS)

First Edition

Silvano Gai
Tommi Salli
Roger Andersson

Silvano Gai

Silvano Gai, who grew up in a small village near Asti, Italy, has over twenty seven years of experience in computer engineering and computer networks. He is the author of several books and technical publications on computer networking and has written multiple Internet Drafts and RFCs. He is responsible for 30 issued patents and 50 patent applications. His background includes seven years as a full professor of Computer Engineering, tenure track, at Politecnico di Torino, Italy and seven years as a researcher at the CNR (Italian National Council for Scientific Research). For the past eleven years, he has been in Silicon Valley where in the position of Cisco Fellow, he was an architect of the Cisco Catalyst family of network switches, of the Cisco MDS family of storage networking switches and of the Nexus family of data center switches. Silvano teaches a course on the topics of this book at Stanford University (http://scpd.stanford.edu/certificates/fcoe).

Tommi Salli

Tommi Salli, who was born and raised in Finland, has close to twenty years of experience working with computers. He has extensive server and application background from companies like SUN Microsystems, VERITAS Software, which later got bought by Symantec from where he moved to Nuova Systems that got bought by Cisco. He has hold different positions from Sales Engineer to Technology Scouting in the office of CTO from Product management to architect and during his journey he has been responsible for 7 patent applications. He has started his career in Finland and for the past five years he has been in Silicon Valley and is currently working for Cisco systems as a Technical Marketing Engineer.

Roger Andersson

Roger Andersson was born in Stockholm, Sweden and has spent over 18 years in the computer industry in both Sweden and the United States. Roger's experience includes over 12 years in the CLARiiON Engineering Division at EMC and 5 years at VERITAS/Symantec where Roger worked as a Technical Product Manager focusing on systems management, server and application automated provisioning. Roger is currently working at Cisco as a Manager, Technical Marketing, where he is focused on the system management aspects of a Unified Computing System.

Project California:
a Data Center Virtualization Server

Unified Computing System (UCS)

First Edition

Silvano Gai
Tommi Salli
Roger Andersson

To my wife Antonella, my daughters Eleonora, Evelina, and my son
Marco

To my wife Sari, my daughter Tara and my son Sean

To my wife Kristine Mitchell

and to our parents for their support in the early journey

Table of Contents

Acknowledments

This book is the result of a collaborative effort inside and outside Cisco®. Many people have contributed and in particular the authors want to express gratitude to:

Cisco® Contributors & Reviewers:

- Billy Moody
- Carlos Pereira
- Christopher Paggen
- Christopher Travis
- Corey Rhoades
- Damien Philip
- Dante Malagrino
- David Lawler
- Dino Farinacci
- Fabio Ingrao
- Fausto Vaninetti
- Garth O'Mara
- Glenn Charest
- Irene Golbery
- John McDonough
- Jose Martinez
- JR Rivers
- Landon Curt Noll
- Liz Stine
- Louis Watta
- Madhu Somu
- Mauricio Arregoces
- Maurizio Portolani
- Michelangelo Mazzola

- Mike Dvorkin
- Mike Galles
- Philip Manela
- Robert Burns
- Susan Kawaguchi
- Tjerk Bijlsma
- Victor Moreno
- Walter Dey

Cisco® Legal and Corporate Identity:

- Donna Helliwell
- Mike Liu
- Suzanne Stout

For Emulex® contribution:

- Ranga Bakthavathsalam

For Intel® contribution:

- Jason Waxman
- Jeff Pishny
- Leslie Xu
- Matthew Taylor
- Richard Tattoli
- Shannon Poulin
- Stephen Thorne
- Sunil Ahluwalia

For QLogic® contribution:

- Manoj Wadekar

Silvano's wife Antonella has patiently reviewed the manuscript hunting for errors. Thank You.

Finally we would like to acknowledge Julie Totora (www.paperumbrella.com) who helped us redraw most of the graphics.

Preface

This book is the result of the work done by the authors initially at Nuova Systems and subsequently at Cisco Systems on Project California, officially known as Cisco Unified Computing Systems (UCS).

Project California is a very broad data center server project that requires different expertise. For this reason, the authors, from three very different backgrounds (and even different countries), have decided to combine their knowledge to publish this book together.

The book describes project California from an educational view: we have tried to provide updated material about all server components and new data center technologies, as well as how these components and technologies are used to build a state of the art data center server.

We wish to express our thanks to the many engineering and marketing people from both Nuova Systems and Cisco Systems with whom we have collaborated in the recent years.

Project California would not have been possible without the determination of the Cisco's management team. They understood the opportunity for Cisco in entering the server market and decided to pursue it. Our gratitude goes out to them.

Nomenclature

Engineering projects are identified from inception to announcement by fantasy names that are often after geographical places to avoid any trademark issue. Project California is not an exception. Project California or simply "California" is the name of the overall systems and city names, such as Palo and Menlo, are used to identify specific components.

Before the launch, Cisco decided to assign project California the official name of "Unified Computing System (UCS)", but the term California will continue to be used informally.

Cisco UCS Manager

The Cisco UCS Manager software integrates the components of the Cisco Unified Computing System into a single, seamless entity. The UCS Manager is described in Chapter 6.

Cisco UCS 6100 Series Fabric Interconnects

The Cisco UCS 6100 Series Fabric Interconnects are a family of line-rate, low-latency, lossless 10 Gigabit Ethernet, Cisco Data Center Ethernet, and Fiber Channel over Ethernet (FCoE) switches, designed to consolidate the I/O at the system level. The Fabric Interconnects are described in Chapter 5.

Cisco UCS 2100 Series Fabric Extenders

The Cisco UCS 2100 Series Fabric Extender provides an extension of the I/O fabric into the blade server enclosure providing a direct 10 Gigabit Cisco Data Center Ethernet connection between the blade servers and the Fabric Interconnects, simplifying diagnostics, cabling, and management. The Fabric Extenders are described in Chapter 5.

Cisco UCS 5100 Series Blade Server Enclosures

The Cisco UCS 5100 Blade Server Enclosures physically house blade servers and up to two fabric extenders. The Blade Server Enclosures are described in Chapter 5.

Cisco UCS B-Series Blade Servers

The Cisco UCS B-Series Blade Servers are designed for compatibility, performance, energy efficiency, large memory footprints, manageability, and unified I/O connectivity. They are based on Intel® Xeon® 5500 series processors (described in Chapter 2.), aka Nehalem-EP. The Blade Servers are described in Chapter 5.

Cisco UCS Adapters

The Cisco UCS Adapters are installed on the UCS B-Series Blade servers in order to provide I/O connectivity through the UCS 2100 Series Fabric Extenders. The different adapters are described in Chapter 4.

Carbon footprint

When we started this book, we discussed the possibility of publishing it either in electronic format or as a regular book. Through our experience we have seen books have a larger impact compared to PDF files distributed on a CD or over the Internet. One of the few disadvantages of printed books is their carbon footprint that, even if minimum, still exists.

The carbon footprint is the total set of GHG (greenhouse gas) emissions caused, either directly or indirectly, by the publication of the book.

Maurizio Portolani, a friend of the authors, suggested a carbon offset strategy, i.e., the mitigation of carbon emissions, through the defense of forests. The authors thought it was a wonderful idea and decided to donate all the royalties of this book to the Orangutan Conservancy.

This charity helps conservation projects that are working to protect the forests and to stop the destructive impact logging has on the environment. These projects also fight poaching and help communities develop long-term plans for alternate sustainable incomes (pictures courtesy of the Orangutan Conservancy).

Using this approach this book has a negative carbon footprint.

1. Introduction

Project California is one of the largest endeavors ever attempted by Cisco®. It is a competitive data center computing solution that is not a simple "me too" design, but a radical paradigm shift.

When we started this book, we were focused on describing project California and pointing to other books for the reference material. Unfortunately, after repeated visits to the Stanford University bookstore and other prominent bookstores in Silicon Valley we were unable to identify any updated reference books. As a consequence we have included this reference data ourselves.

The result is a book that consist of 50% reference material applicable to any server architecture and 50% specific to the California project. The reference material includes updated processor, memory, I/O, and virtualization architectures.

Project California has a large ecosystem of partners, several of which have provided material for this book. We thank Intel®, Emulex®, Qlogic®, and BMC® for their contributions and help in making this book a reality.

With any book there is always the question of how much time is spent in proofreading it and making it error free. Our management answered this for us saying that the book had to be available at the time of California launch. This has put a hard limit on the spell checking and improvement on graphics. If you find any error, or you have any advice for improvement, please email them to ca-book@ip6.com.

Finally, this book is neither a manual, nor a standard, nor a release note, nor a product announcement. This book was written to explain the concepts behind project California to a large audience. It is not authoritative on anything; please reference the appropriate official documents when evaluating products, designing solutions or conducting business: the authors do not provide any guarantee of the correctness of the content and are not liable for any mistake or imprecision.

1.1. Data center challenges

Data centers are the heartbeat of large corporations IT infrastructures. A typical Fortune 500 company runs thousands of applications worldwide, stores petabytes (10^{15}) of data and has multiple data centers along with a disaster recovery plan in place. However this huge scale infrastructure often comes at a huge cost! Data centers require expensive real estate, they use a lot of power and in general they are expensive to operate.

To have a better understanding of how large these new data centers can become, reference [35] contains examples like:

- Google® The Dalles (OR) facility 68,680 Sq Ft (6,380 m²);

- Microsoft®Quincy, 470,000 Sq Ft (43,655 m²), 47 MWatts;

- Yahoo® Wenatchee & Quincy, 2 Million Sq Ft (185,800 m²);

- Terremark® – NOTA, 750,000 Sq Ft (68,6767 m²), 100MWatts.

These first few sections will analyze these issues in more detail.

1.1.1. Environmental Concerns - "Green"

You've probably noticed that green is everywhere these days: in the news, politics, fashion, and technology; data centers are no exception. The U.S. Environmental Protection Agency and the U.S. Department of Energy have a joint program to help businesses save money while protecting the environment through energy efficient products: it is called ENERGY STAR [40]. The money saving can be estimated by checking the average retail price of electricity [42].

ENERGY STAR is a proven energy management strategy that helps measure current energy performance, setting goals, tracking savings, and rewarding improvements. For example, the ENERGY STAR 4.0 standard which took effect in July 2007, requires power supplies in desktops, laptops and workstations, to be 80% efficient for their load range. In addition, it places limits on the energy used by inactive devices, and requires systems to be shipped with power management features enabled.

Another effort, called Climate Savers® smart computing [41], started in the spirit of WWF's Climate Savers program. It has mobilized over a dozen companies since 1999 to cut CO_2 emissions, demonstrating that reducing emissions is part of a good business. Their mission is to reduce global CO_2 emissions from computer operation by 54 million tons per year, by 2010. This is equivalent to the annual output of 11 million cars or 10–20 coal-fired power plants.

Participating manufacturers commit to producing products that meet specified power-efficiency targets, and members commit to purchasing power-efficient computing products.

Climate Savers Computing Initiative starts with the ENERGY STAR 4.0

specification for desktops, laptops and workstation computers and gradually increases the efficiency requirements in the period 2007 - 2011.

Project California will be compliant with Energy Star and is designed to meet the power-efficiency targets of Climate Savers from day one.

1.1.2. Server Consolidation

In trying to save energy it is always important to have a balanced approach between "every watt counts" and "let's target first the big consumers".

Speaking about the former, I remember a friend of mine telling me that he was drilling holes in the handle of his toothbrush before leaving for a long hiking trip, in order to reduce as much as possible the weight to carry.

The equivalent is to try to optimize as much as possible any component that uses power in a datacenter. For example, the networks elements consume approximately 14% of the overall power. By making them 50% more efficient (a goal very difficult to achieve) we will save 7% of the power.

Speaking about the latter, servers are the greatest power consumers in the data center and often most of the servers are very lightly loaded. Some statistics speak of an overall server utilization ranging from 5% to 10%. Let's assume that the servers use 70% of the power and that we reduce their number of a factor of five: load will increase to 25 – 50% and power will be reduced to 14%, with a net saving of 56%.

The two techniques used in combination will produce even larger results. The key technology for reducing the number of servers is "server virtualization".

1.1.3. Virtualization

Virtualization is a broad and overloaded term that refers to the abstraction of computer and network resources. For example VLAN (Virtual LAN) and VSAN (Virtual SAN) are forms of network virtualization.

What is becoming increasingly important for data centers is server virtualization, in particular hardware-assisted server virtualization. Wikipedia® defines it as: "*Hardware-assisted virtualization is a virtualization approach that enables efficient full virtualization using help from hardware capabilities, primarily from the host processors.*" Full virtualization is used to simulate a complete hardware environment, or virtual machine, in which an unmodified "guest" operating system (using the same instruction set as the host machine) executes in complete isolation. Hardware-assisted virtualization was first implemented on the IBM® System/370®, and was recently (2007) added to x86 processors (Intel VT® or AMD-V®).

According to Gartner, "*Virtualization is the highest impact trend changing infrastructure and operations through 2012. It will change how you manage, how and what you buy, how you deploy, how you plan, and how you charge [36]*". Several studies by the research firm IDC support this claim. The firm reports 22 percent of servers today as being virtualized and expects that number to grow to 45 percent over the next 12 to 18 months [37]. Another IDC study predicts that the number of logical servers generated on virtualized servers will surpass the number of non-virtualized physical server units by 2010 [38].

Examples solutions for virtualization in X86 based processor systems include VMware® ESX®, Microsoft Hyper-V® and Linux® Xen®.

With efficient hardware-assisted virtualization, multiple low usage servers can be virtualized (i.e., transformed into a Virtual Machine) and multiple VMs can be run simultaneously on the same physical server. VMs can also be moved from one server to another for load balancing or disaster recovery.

Organizations worldwide are already beginning to take advantage of this model. The 2007 IDC study, for example, showed that 50% of all VMware ESX® users had adopted VMotion capability [39]. This technology enables live migration – moving guests from one physical server to another with no impact to end users' experience. By giving IT managers the ability to move guests on the fly, live migrations make it easier to balance workloads and manage planned and unplanned downtimes more efficiently.

Project California, with its large memory footprint and native virtualization support, may help cut the number of servers by one order of magnitude, thus achieving an even larger power saving.

1.1.4. Real Estate Power and Cooling

In building a new data center the number of servers that can be installed per square foot (square meter) is the result of many considerations, a primary one being how much cooling can be provided. In fact all the power consumed by the data center equipment is transformed into heat that needs to be removed by the air conditioning systems.

Current data centers range from 50 to 200 Watts per square foot (500 to 2,000 W/m^2), which is the sweet spot for the current cooling technologies. New data centers are designed in the 300 to 500 Watts per square foot (3 to 5 KW/m²), but they require expensive cooling technologies that must be uninterruptible, since with such a high power load the rise of temperature in the case of a fault in the cooling system can be as high as 25° F (14° C) per minute.

The Watts per square foot map directly into Watts per rack. Assuming that to install a rack every 40-60 square feet (4-6 m²), current data centers have 2 to 10 KW/rack and future design may go as high as 12-25 KW/rack [35].

1.1.5. Cabling

Cabling is another big topic in data center designs. A mix of copper cables and fiber optics has been used till now, with two major topologies named Top of Rack (ToR) and End of Row (EoR).

The EoR approach places the network equipment (mainly LAN and SAN switches) at the end of the row and utilizes switches with a higher port count (128 – 512 ports) in the attempt to reduce the number of network devices that need to be managed. This implies longer cables with different lengths from the servers to the networking equipment. At 1 Gbps these cable may still be copper, but at 10 Gbps fiber may be the only solution.

While a fiber cable may be cheaper than a copper cable, when installation and transceiver are considered (a fiber connection cost may exceed $3,000), if two or four links are used per server, this will exceed the cost of the server, without considering the cost of the switch ports.

In contrast, the ToR approach places the networking equipment as close as possible to the servers, typically at the top of each rack or every few racks. ToR switches are typically low port count (26 – 52 ports) fixed configuration switches. Distance from the servers to the ToR switches are limited typically to 33 feet (10 meters) and copper cables can easily be deployed, even with 10Gbps speed, achieving an all copper cabling within a rack. Fibers are used to connect the ToR switches to the central network equipment (e.g., switches and Fibre Channel directors). All the servers connected to the same ToR switch share the cost of these fibers. This approach has more management points than the previous one.

In both cases, cabling is a significant portion of the CAPEX (capital expenditure) of a data center and it also tends to restrict airflow in the rack and under the floor, negatively impacting cooling effectiveness (OPEX: operating expenditure).

In the past, three different parallel networks were deployed in the data center: Ethernet for LAN traffic, Fibre Channel for SAN traffic, and a separate Ethernet LAN for management. Sometimes additional dedicated networks have also been deployed, for example a separate Ethernet or Fibre Channel network for backups and Infiniband for High Performance Computing. In future data centers Ethernet is emerging as the unified network. Terms like Unified Fabric, I/O consolidation and FCoE (Fibre Channel over Ethernet) are used to indicate the adoption of Ethernet as the sole network in the data center. This of course greatly simplifies cabling.

California is a system designed according to the ToR approach, all the servers that compose a California system are connected to two Fabric Interconnects that are placed at the top of one or every few racks and they communicate by using a Unified Fabric approach, thus greatly minimizing the cabling.

1.1.6. Disaster Recovery

Disaster strikes unexpectedly and large organizations must have a disaster recovery plan in place. This plan may be required by law to comply with government regulations. Data centers are no exception and they must be replicated at safe distances to minimize the possibility that both data centers are affected by the same disaster. This implies replicating storage and servers and being able to restart the computing services on the backup data center in a minimal amount of time, known as Recovery Time Objective (RTO).

Virtualization is again our friend, since it allows moving Virtual Machines (VMs) from one site to another. Often VM movement is possible only inside the same layer 2 network (e.g., VLAN) since the MAC and IP addresses of the virtual server are part of the VM and in most cases they must be preserved during the movement.

VMs have created a renewed interest in layer 2 networks which at one point seemed to be losing ground to layer 3 networks, e.g., IP routers. Inside the data center this will require replacing the spanning tree protocol with a more efficient layer 2 multipath approach. The IETF TRILL project [18] is a good example of this evolution: it proposes IS-IS in conjunction with MAC-in-MAC encapsulation to enable layer 2 multipathing. Efficient techniques to extend layer 2 networks across two or more data centers must also exist and we discuss these in the following section on Network Virtualization.

Finally, a simple way to move server configurations from one data center to another must exist. In a California system the identity of the various components

(MAC, addresses, UUIDs, WWNs, etc.) are not burned in the hardware, but contained in a configuration file. This technique allows the administrator to recreate an identical California system in a disaster recovery site by simply moving the configuration file.

1.1.7. Network Virtualization

The role of the network in the virtualization space should address two complementary aspects: first, the use of network functionality in support of virtualized compute environments, and second the virtualization of the network elements.

When planning for disaster recovery or managing workloads across multiple data center facilities, larger and pervasive Layer 2 networks provide a multitude of operational advantages.

Larger and multi-site layer 2 networks should not create an unnecessary operational burden and should maintains the scalability and stability provided today by IP networks.

A promising solution is "MAC routing", which enables VPN solutions in which Layer 2 connectivity can be provided between separate Layer 2 domains, while preserving all the benefits of an IP based interconnection.

In MAC routing, Layer 2 reachability information is distributed in a control protocol much like that used in a Layer 3 network. This protocol learning is the cornerstone to maintaining the failure containment and loop-free path diversity characteristics of an IP network while still providing Layer 2 connectivity. There are techniques in MAC routing to ensure that reconvergence events, broadcasts and unknown unicast flooding can be localized and kept from propagating to multiple data centers.

In the MAC routing model, traffic forwarding is done natively in IP, which makes the solution transport agnostic. The network architect now has the freedom of leveraging any Layer 1, Layer 2 or Layer 3 service between their data centers. The key here is that the operational model doesn't change from one transport to another. Thus, there aren't any complex interactions between the transport core and the overlaid Layer 2 extensions. This is a big departure from the transport restrictions imposed by current label switched VPN technologies. In short, MAC routing is totally transparent to the core and therefore minimizes the impact on design and operations that the extension of Layer 2 may impose on the network.

The Cisco Nexus 7000 brings MAC routing functionality at the L2/L3 boundary, maintaining the benefits of L3 routing between Layer 2 domains that may require Layer 2 connectivity. The Layer 3 intelligence may be realized in the underlying IP core or on the overlay control plane itself, allowing inter-domain Layer 2 traffic to inherit the wide variety of enhancements traditionally seen in a Layer 3 network. Some examples include: Loop Free Multi-pathing enabling load distribution for a high bandwidth interconnect, optimal Multicast replication, single control plane for multicast and unicast, Fast Re-route with Equal Cost Multi-Path routing.

MAC routing enables Layer 2 VPN solutions, which by extension enable Layer 3 VPN solutions. These are used more and more in order to virtualize the network itself and realize the consolidation benefits of a virtual network environment.

1.1.8. Desktop Virtualization

An important form of virtualization is desktop virtualization. Many large corporations have thousands of desktops distributed in multiple sites that are complex to install and maintain. The alternative has been to provide a terminal server service, but this deprives the users from a full PC desktop experience.

Desktop virtualization or Virtual Desktop Infrastructure (VDI) proposes to give the system administrator and the end users the best of both worlds, i.e., a full PC experience for the users that is centrally provisioned and managed by the systems administrators.

With desktop virtualization the system managers can provide new applications, upgrade existing ones and upgrade or patch the operating systems of all the desktops in a centralized manner. Data is also stored in a consistent way and can be backed-up and restored centrally.

Companies like VMware®, Microsoft®, and Citrix® have proposed a few different approaches. They all have services running in a data center, often on virtual machines, and different kinds of desktops that range from classical PCs, to laptops, to thin clients. They vary in where the applications are run, the degree of flexibility they provide to the user, how anonymous the desktops are, and the technologies used to install applications and to communicate between the desktops and the servers.

The Remote Desktop Protocol (RDP) has been used in many installations, but other solutions are appearing on the market, but a clearly winning architecture has not yet emerged.

Project California is a platform that can run thousands of virtual machines on a single system and it is therefore perfectly suited to host large desktop virtualization environments.

1.1.9. Cloud Computing

Cloud computing is a popular buzzword used to indicate a more economic way to run applications. The idea is to use a "cloud" composed of commodity servers, storage and networking equipment. The cloud runs applications on demand, through a flexible management system capable of growing or shrinking resources applications need [43], [44]. Cloud computing brings a huge amount of flexibility as the administrator can scale up or down his/her compute as and when he/she needs to.

With cloud computing, applications can be accessed from anywhere, at anytime. Clouds may be private to an enterprise or public. The most well known example of public cloud computing is probably Amazon® EC2®.

Of course, not all applications are suited for public clouds. Significant concerns exist in relation to data security. Regulatory requirements can mandate keeping multiple copies of data in different geographical locations. Other concerns include latency and application availability.

For these reasons, many customers are evaluating the possibility to create private clouds inside their enterprises as a way to build a cost-effective data center infrastructure, suitable for on-demand application deployment. This should significantly increase server utilization.

Moreover, customers are starting to require servers that are "anonymous", i.e., that can be easily repurposed. In addition, these new servers must be capable of supporting a high number of virtual machines, of easily moving them from one server to another, and have a management system which is policy-driven and provides a reach API interface (to interface with the cloud software).

Cisco California is a server designed with cloud computing in mind.

1.2. Evolution of data centers

This section describes the evolution of data center server architectures.

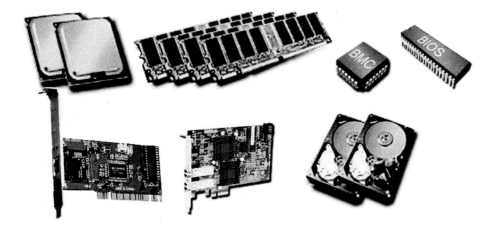

Figure 1: Server components

1.2.1. Stand-alone servers

Also known as discrete servers, they are independent servers that have a proprietary form factor. They may range from a desk-side PC to a large mainframe. They all have the basic components shown in Figure 1, but they may have a different number of processors, amount of memory, I/O capacity, expansion slots, and integrated storage.

The largest ones run multiple different applications and support a large number of users. Often they use some form of virtualization software that allows running multiple operating systems at the same time. However, no matter how powerful a discrete server is, it always has scalability limits and it provides an operating environment that may not be optimal for all applications.

To overcome these scalability limitations two different approaches have been used: scale-up and scale-out.

1.2.2. Scale-up

Scale-up, aka scale vertically, is a term used to indicate a scaling strategy based on increasing the computing power of a single server by adding resources in term of more processors, memory, I/O devices, etc.

Such vertical scaling of existing systems also enables them to leverage virtualization technology more effectively, as it provides more resources for the hosted operating systems and applications.

Scale-up limits the number of management points, potentially easing security policy enforcement.

While attractive for some applications, most data centers prefer to use standard components and adopt a scale-out approach.

1.2.3. Scale-out

Scale-out, aka scale horizontally, is a term used to indicate a scaling strategy based on increasing the number of servers. It has the big advantage that the administrator can more easily re-purpose compute as required.

Scale-out strategy are commonly adopted in association with Intel X86 servers. These type of servers, based on the PC architecture, have seen in the recent years a continued price drop and performance increase. These "commodity" systems have now reached a computing capacity sufficient to run the majority of applications present in data centers. They may also be interconnected in clusters to perform HPC (High Performance Computing) applications in scientific fields like modeling and simulation, oil and gas, seismic analysis and biotechnology, previously possible only on mainframes or supercomputers.

The scale-out model has created an increased demand for shared data storage with very high I/O performance, especially where processing of large amounts of data is required, such as for databases.

1.2.4. Scale-up vs. scale-out

There are trade-offs between the two models.

Scale-up requires dedicated and more expensive hardware and provides a limited number of operating system environments, and there are limitations on the amount of total load that a server can support. It has the advantage of having few management points and a good utilization of resources and therefore it tends to be more power and cooling efficient than the scale-out model.

Scale-out dedicates each server to a specific application. Each application is limited by the capacity of a single node, but each server can run the most appropriate OS for the application with the appropriate patches applied. There is no application interference and application performance is very deterministic. This approach clearly explodes the number of servers and it increases the management complexity.

1.2.5.   Rack-optimized servers

With scale-out constantly increasing the number of servers, the need to optimize their size, airflow, connections, and to rationalize installation becomes obvious.

Rack-optimized servers are the first attempt to solve this problem (see Figure 2). Also known as rack-mounted servers, they fit in 19-inch wide racks and their height is specified in term of Rack Unit (RU), one rack unit being 1.75 in (44.45 mm) high. A typical Intel-based server fits in one RU and it dissipates approximately 500 Watts. The racks are typically 42 RUs, but it is normally impossible to provide enough power and cooling to fill an entire rack with servers.

Typical Data Centers have 5 to 10 KW of power and cooling available per rack and therefore 10 to 20 servers are installed in a rack. The remaining space is filled with patch panels. Sometimes ToR (Top of the Rack) switches are installed to aggregate the traffic generated by the servers present in few adjacent racks. In other designs larger EoR (End of Row) switches are used to connect all the servers in a row and hybrid schemes ToR/EoR are also possible.

The benefits of this approach are rational space utilization and high flexibility: the relative large server size allows adopting the latest processor with a larger memory, and several I/O slots. The weaknesses are in the lack of cabling rationalization, in the serviceability that is not easy, and in the lack of power and cooling efficiency, since each server has its own power supplies and fans.

Figure 2: Rack-mounted servers

Rack-mounted servers are just a simple repackaging of conventional servers in a form factor that allows installing more servers per square foot of data center floor, but without significant differences in functionality.

At the time of writing rack-optimized servers account for approximately 50% of the overall server markets.

1.2.6. Blade-Server

Blade-servers were introduced as a way to optimize cabling and power efficiency of servers compared to rack mounted. Blade-server chassis are 6 to 12 RUs and can host 6 to 14 computing blades, plus a variety of I/O modules, power supplies, fans, and chassis management CPUs (see Figure 3).

Blade-server advantages are shared chassis infrastructure (mainly power and cooling), rationalization in cabling, and the capability to monitor the shared infrastructure. The number of management points is reduced from one per server to one per chassis, but the chassis is often an added artificial point of aggregation. The concept of chassis is not of primary importance, for example, when defining a pool of servers to be used for a particular application or for virtualization.

At the time of writing blade-servers account for approximately 10% of the overall server markets, but their adoption is accelerating and this percentage is growing rapidly.

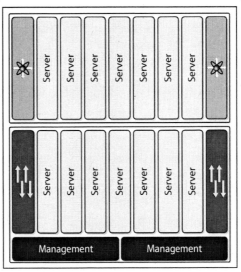

Figure 3: Blade Servers

1.2.7. Server Sprawl

Today most servers run a single OS (commonly some form of Windows® or Linux®) and a single application per server (see Figure 4). This deployment model leads to "Server Sprawl", i.e., a constantly increasing number of servers with an extremely low CPU utilization, as low as 5% to 10% average utilization. This implies a lot of waste in space, power and cooling.

The benefits of this deployment model are isolation (each application has guaranteed resources), flexibility (almost any OS/application can be started on any server), and simplicity (each application has a dedicated server with the most appropriate OS version). Each server is a managed object and since each server runs a single application, each application is a managed object.

This architecture allows the consistent setting of differentiated policies for the different applications. The network has one (or more) physical port for each application and on that port it can ensure QoS, ACLs, security, etc. This works independently from the fact that the switches are inside the blade-server or outside.

Nonetheless Server Sprawl is becoming rapidly unacceptable due to the waste of space, power, and cooling. In addition, the management of all theses servers is an administrator nightmare as well very expensive.

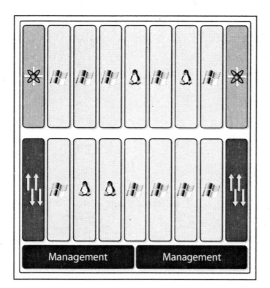

Figure 4: Single OS/application per server

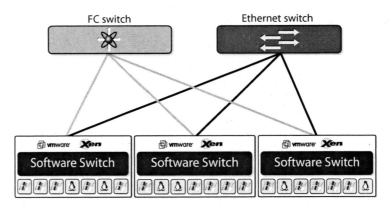

Figure 5: Server Virtualization

1.2.8. *Virtualization*

Virtualization is a technology used to reduce server sprawl. As opposed to the chassis being a container, with virtualization a server is a container of multiple logical servers (see Figure 5 and Figure 6). The virtualization software contains a software switch that is equivalent to the switch present in a chassis. The advantages of server virtualization are utilization, mobility, and availability. The disadvantages are lack of distributed policies, security, diagnostics and performance predictability.

At the time of printing all Data Centers are evaluating virtualization, but al-

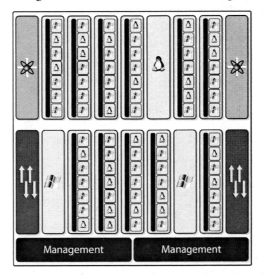

Figure 6: Virtualization deployed on a blade server

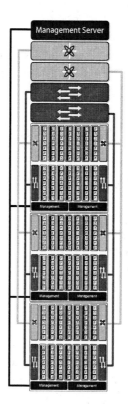

Figure 7: Blade servers interconnected by external switches

most none are deploying critical applications on virtualization, due to the previous limitations.

1.2.9. Server Deployment Today

All of the server evolution described up to now has mainly been "an evolution of size" not a significant change in the model. Scale-up means larger servers, scale-out means more servers and therefore more network infrastructure. For example, Figure 7 shows three blade servers, running virtualization, mounted in a rack and interconnected by external switches.

Management tools are very often an afterthought: they are applied to servers, not deeply integrated with them. This causes a growth in the number of tools required to manage the same server set and in general it is difficult to maintain policy coherence, to secure and to scale.

1.3. Unified Computing System (UCS)

The Cisco Unified Computing System (UCS), aka Project California, is an attempt to eliminate, or at least reduce, the limitations present in current server deployments.

Cisco's definition for Unified Computing is:

> *Unified Computing unifies network virtualization, storage virtualization, and server virtualization into one, within open industry standard technologies and with the network as the platform.* "

California is a component of the Cisco Data Center 3.0 architecture. It is a scalable compute platform based on the natural aggregation point that exists in any Data Center: the network.

California is composed of fabric interconnects that aggregate a set of blade chassis. Compared to blade servers it is an innovative architecture. It removes unnecessary switches, adapters and management modules, i.e., it has ⅓ less infrastructure compared to classical blade-servers. This results in less power and cooling and fewer management points, that leads to an increased reliability.

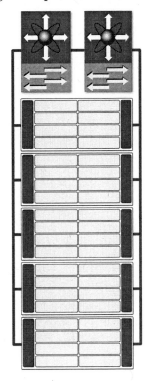

Figure 8: A UCS (aka California) system

For example, Figure 8 shows a California system with 40 blades. When compared with the three blade servers of Figure 7 (42 blades total) six Ethernet switches, six Fibre Channel Switches, and six management modules are no longer needed. This equates in power and cooling saving, but also in a reduced number of management points.

The saving grows larger with a larger number of blades. A single California system may grow as large as 320 blades. Figure 9 shows a single California system with 280 blades, hosted in 7 racks.

The most important technology innovations introduced in California are:

- Embedded management: California does not require a separate management server, since it has an embedded management processor that has global visibility on all the elements that constitute a California system. This guarantees coordinated control and provides integrated management and diagnostics.

- Unified Fabric: California is the first server completely designed around the concept of Unified Fabric. This is important since different applications have different I/O requirements. Without Unified Fabric it is difficult to move applications from one server to another while also preserving the appropriate I/O requirement. Unified Fabric not only covers LAN, SAN, and HPC, but also the management network.

- Optimized Virtualized Environment: designed to support a large

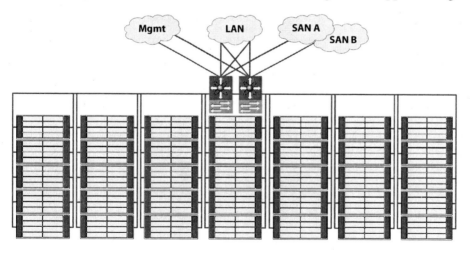

Figure 9: A single California system

number of virtual machines not only from the point of view of scale, but also from policy enforcement, mobility, control, and diagnostics. This is achieved by working simultaneously on the three axis of Figure 10, i.e., by using state of the art CPUs, expanding the memory, and adopting sophisticated I/O techniques.

- Fewer servers with more memory: the California computing blades using Cisco Memory Expansion Technology can support up to four times more memory when compared to competing servers with the same processor. More memory enables faster database servers and more virtual machines per server, thus increasing CPU utilization. Fewer CPUs with more utilization lower the cost of software licenses.

- Stateless Computing: server attributes are no longer tied to physical hardware which guarantees seamless server mobility. The blades and the blade enclosures are completely stateless. California also puts a lot of attention on network boot (LAN or SAN) with the boot order configurable as a part of the service profile.

Since the management is embedded, there is no need to have a separate server for the management software with the required hardware configuration and OS patches, and to have complex installation procedures that are prone to error and that can cause system downtime.

California is not only a single management point, it is a single system. All

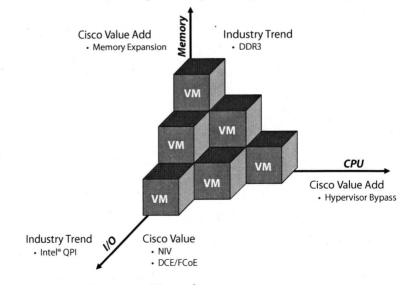

Figure 10: Virtualization improvements

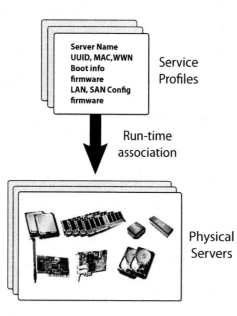

Figure 11: Service profile

California features are made available to any computing blade that is added. California has a comprehensive model driven architecture, in which all components are designed to integrate together.

At the core of the California architecture is the concept of a "Service Profile", i.e., a template that contains the complete definition of a service including server, network, and storage attributes. Physical servers in California are totally anonymous and they get a personality when a service profile is instantiated on them (see Figure 11).

This greatly simplifies service movement, since a service profile can be simply moved from one blade in one chassis (see Figure 12, Chassis-1/Blade-5) to another blade in another chassis (Chassis-9/Blade 2). The uniform I/O connectivity based on Unified Fabric makes this operation straightforward.

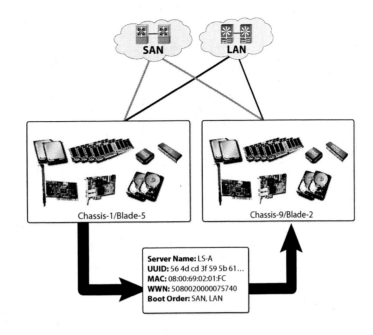

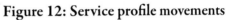

Figure 12: Service profile movements

2.   Server architectures

Processor, memory and I/O are the three most important subsystems in a server from a performance perspective. At any given point in time one of them tends to become a bottleneck from the performance perspective. We often hear of applications that are CPU-bound, memory-bound or I/O-bound.

In this chapter we will examine these three subsystems with particular reference to servers built according to the IA-32 (Intel® Architecture, 32-bit), often generically called x86 architecture. In particular we will describe the last generation of Intel® processors compatible with the IA-32 architecture, i.e., the Intel® microarchitecture (formerly known by the codename Nehalem)[1]. The first processor of this family is the Core i7 for desktops. The Intel® Xeon® 5500 is another member of this family designed for server applications. Project California in its first embodiment uses the Intel® Xeon® Processor 5500 series.

2.1.   The processor Evolution

Modern processors or CPUs (Central Processing Units) are built using the latest silicon technology and they pack millions of transistors and megabytes of memory on a single die (blocks of semiconducting material that contains a processor).

Multiple dies are fabricated together in a silicon wafer; each die is cut out individually, tested, and assembled in a ceramic packaging. This involves mounting the die, connecting the die pads to the pins on the package, and sealing the die.

At this point the processor in its package is ready to be sold and mounted on servers. Figure 13 shows a packaged Intel® Xeon® 5500.

2.1.1.   Sockets

Processors are installed on the motherboard using a mounting/interconnection structure known as a "socket". Figure 14 shows a socket used for an Intel® Processor.

The number of sockets present on a server motherboard determines how many processors can be installed. Originally servers had a single socket, but more re-

---

1   The authors are thankful to Intel® Corporation for the information and the material provided to edit this chapter. Most of the pictures are courtesy of Intel® Corporation.

Figure 13: An Intel® Xeon® 5500 Processor

cently, to increase server performance, 2, 4 and 8 socket servers have appeared on the market.

In the evolution of processor architecture, for a long period of time performance improvements were strictly related to clock frequency increases. The higher the clock frequency, the shorter the time it takes to make a computation, and therefore the higher the performance.

As clock frequencies approached a few GHz, it became apparent the physics involved would limit further improvement in this area. Therefore, alternative

Figure 14: An Intel® Processor Socket

ways to increase performance had to be identified.

2.1.2. Cores

The constant shrinking of the transistor size (today at 45 nanometers) has allowed the integration of millions of transistors on a single die: one way to use this abundance is to replicate the basic CPU (the "core") multiple times on the same die.

Multi-core processors (see Figure 15) are now common in the market. They contain, in a single socket, multiple CPU cores (2. 4, 8 are typical numbers). Each core is associated with a level 1 (L1) cache. Caches are small fast memories used to reduce the average time to access the main memory The cores generally share a larger level 2 (L2) cache, the bus interface and, in general, the external die connections.

Therefore, in modern servers the number of cores is the product of the number of sockets times the number of cores per socket. For example, servers based on Intel® Xeon® Processor 5500 Series (Nehalem-EP) typically use two sockets and four cores per sockets for a total of eight cores.

Figure 16 shows a more detailed view of a dual-core processor. The CPU's main components (instruction fetching, decoding and execution) are duplicated, but the access to the system busses is common.

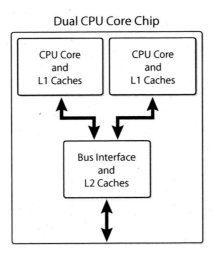

Figure 15: Two CPU cores in a socket

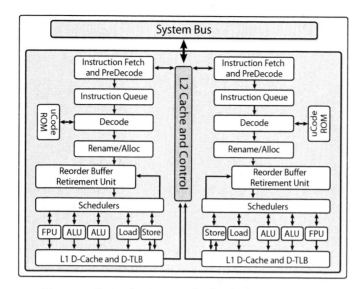

Figure 16: Architecture of a dual-core processor

2.1.3. Threads

To better understand the implication of multi-core architecture, let's consider how programs are executed. A server will run a kernel (e.g. Linux, Windows®) and multiple processes. Each process can be further subdivided into "threads". Threads are the minimum unit of work allocation to cores. A thread needs to execute on a single core, and it cannot be further partitioned among multiple cores (see Figure 17).

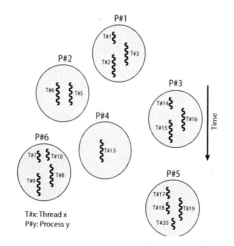

Figure 17: Processes and Threads

Processes can be single-threaded or multi-threaded. A single thread process can execute in only one core, therefore its performance is limited by the performance of that core. A multi-threaded process can execute on multiple cores at the same time and therefore its performance can exceed the performance of a single core.

Since many applications are single threaded, a multi-socket, multi-core architecture is typically convenient in an environment where multiple processes are present. This is always true in a virtualized environment, where a hypervisor allows consolidating multiple logical servers into a single physical server creating an environment with multiple processes and multiple threads.

2.1.4. Intel® Hyper-Threading Technology

While it is true that a single thread cannot be split between two cores, some modern processors allow running two threads on the same core at the same time. This is beneficial since a core has multiple execution units capable of working in parallel and it is difficult for a single thread to keep all the execution units (resources) busy.

Figure 18 shows how the Intel® Hyper-Threading Technology works. Two threads execute at the same time on the same core and they use different resources thus increasing the throughput.

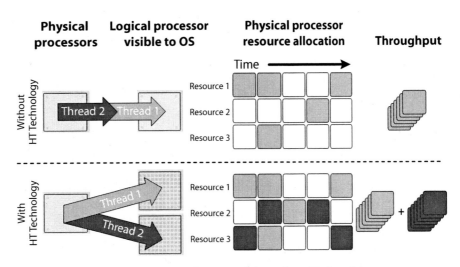

Figure 18: Intel® Hyper-Threading Technology

2.1.5. Front Side Bus

In the presence of multi-sockets and multi-cores it is important to understand how the memory is accessed and how communication between two different cores work.

Figure 19 shows the architecture used by many current generation Intel® processors, known as FSB (Front-Side Bus). In this architecture all traffic is sent across the FSB, a single shared bidirectional bus. In modern processors, this is a 64-bit wide bus that is operated at 4X the bus clock speed and in certain products is operated at an information transfer rate of up to 1.6 GT/s (Giga Transactions per second, i.e, 12.8 GB/s).

The FSB is connected to all the processors and to the chipset, more in particular to the Northbridge (aka MCH: Memory Controller Hub). The Northbridge connects the memory that is shared across all the processors.

One of the advantages of this architecture is that each processor sees all the memory accesses of all the other processors and therefore can implement a cache coherency algorithm to keep its internal caches in synch with the external memory, and therefore with the caches of all other processors.

Platforms designed in this manner have to contend with the shared nature of the bus. As signaling speeds on the bus increase, it becomes more difficult to implement and connect the desired number of devices. Also, as processor and chipset performance increased, in general the traffic flowing on the FSB also

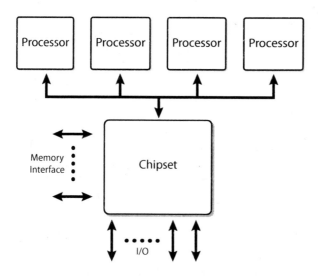

Figure 19: A server platform based on a Front Side Bus

increased, resulting in increased congestion on that shared resource.

2.1.6. Dual Independent Buses

To further increase the bandwidth, the single shared bus evolved into the Dual Independent Buses (DIB) architecture depicted in Figure 20, that essentially doubles the available bandwidth.

However, with two buses all the cache consistency traffic has to be broadcasted on both buses, thus reducing the effective bandwidth. To minimize this problem, "snoop filters" are employed in the chipset to reduce the bandwidth loading.

When a cache miss occurs, a snoop is put on the FSB of the originating processor. The snoop filter intercepts the snoop, and determines if it needs to pass along the snoop to the other FSB. If the read request is satisfied with the other processor on the same FSB, the snoop filter access is cancelled. If the other processor on the same FSB does not satisfy the read request, the snoop filter determines the next action. If the read request misses the snoop filter, data is returned directly from memory. If the snoop filter indicates that the target cache line of the request could exist on the other FSB, the snoop filter will reflect the snoop across to the other segment. If the other segment still has the cache line, it is routed to the requesting FSB. If the other segment no longer owns the target cache line, data is returned from memory. Because the proto-

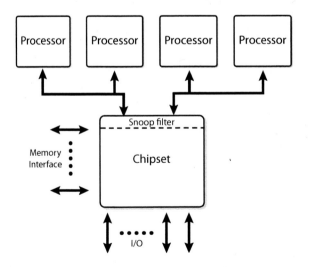

Figure 20: A server platform based on Dual Independent Buses

col is write-invalidate, write requests must always be propagated to any FSB that has a copy of the cache line in question.

2.1.7. Dedicated High Speed Interconnect

The next step of the evolution after the dual independent buses architecture was the introduction of Dedicated High Speed Interconnects (DHSI) as shown in Figure 21

DHSI-based platforms use four FSBs, one for each processor in the platform. Snoop filters are employed to achieve good bandwidth scaling.

Platforms designed using this approach still must deal with the electrical signalling challenges of the fast FSB. They also drive up the pin count on the chipset and require extensive PCB routing to establish all these connections using the wide FSBs.

2.1.8. Intel® QuickPath Interconnect

Beginning with the introduction of the Intel Core i7 processor, a new system architecture has been adopted for many Intel products. This is known as the Intel® QuickPath Interconnect (Intel® QPI). This architecture utilizes multiple high speed uni-directional links interconnecting the processors and chipsets. With this architecture there is also the realization that:

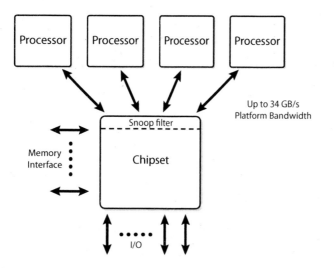

Figure 21: A server platform based on DHSI

- a common memory controller for multiple sockets and multiple cores is a bottleneck;

- introducing multiple distributed memory controllers would best match the memory needs of multi-core processors;

- in most cases having a memory controller integrated into the processor package would boost performance;

- providing effective methods to deal with the coherency issues of multi-socket systems is vital to enabling larger scale systems.

Figure 22 gives an example functional diagram of a processor with multiple cores, an integrated memory controller, and multiple Intel® QPI links to other system resources.

In this architecture all cores inside a socket share IMCs (Integrated Memory Controllers) that may have multiple memory interfaces (i.e, memory buses). IMCs and cores in different sockets talk to each other using the Intel® QPI.

Processors implementing Intel® QPI also have full access to the memory of every other processor, while maintaining cache coherency. This architecture is also called "cache-coherent NUMA (Non-Uniform Memory Architecture)", i.e., the memory interconnection system guarantees that the memory and all the potentially cached copies are always coherent.

Intel® QPI is a point-to-point interconnection and messaging scheme. It uses a point-to-point differential current-mode signaling. On current implementa-

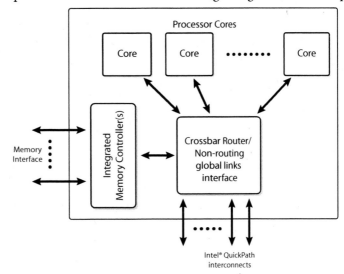

Figure 22: Processor with Intel® QPI

tion each link is composed of 20 lanes per direction capable of up to 25.6 GB/s or 6.4 GT/s (Giga Transactions/second), see Section 2.4.1.

Intel® QPI uses point-to-point links and therefore requires an internal crossbar router in the socket (see Figure 22) to provide global memory reachability. This route-through capability allows one to build systems without requiring a fully connected topology.

Figure 23 shows a configuration of four Intel® Nehalem EX (an evolution of the Intel Xeon 5500 (Nehalem-EP)), each processor has four QPI and interconnects with the three other processors and with the chipsets.

2.2. The memory subsystem

The electronic industry has put a significant effort into manufacturing memory subsystems capable of keeping up with the low access time required by modern processors and the high capacity required by today's applications.

Before proceeding with the explanation of current memory subsystems, it is important to introduce a glossary of the most commonly used terms:

- RAM (Random Access Memory);
- SRAM (Static RAM);
- DRAM (Dynamic RAM);

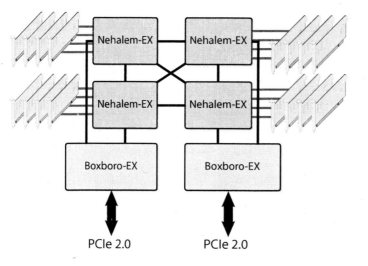

Figure 23: Four-socket Nehalem EX

- SDRAM (Synchronous DRAM);
- SIMM (Single Inline Memory Module);
- DIMM (Dual Inline Memory Module);
- UDIMM (Unbuffered DIMM);
- RDIMM (Registered DIMM);
- DDR (Double Data Rate SDRAM);
- DDR2 (DDR two);
- DDR3 (DDR three).

In particular, the Joint Electron Device Engineering Council (JEDEC) is the semiconductor engineering standardization body that has been active in this field. JEDEC Standard 21 [21], [22] specifies semiconductor memories from the 256 bits SRAM to the latest DDR3 modules.

The memory subsystem of modern servers is composed of RAMs (Random Access Memories), i.e., integrated circuits (aka ICs or chips) that allow the data to be accessed in any order, in a constant time, regardless of its physical location. RAMs can be static or dynamic [27], [28], [29], [30].

2.2.1. SRAMs

SRAMs (Static RAMs) are generally very fast, but smaller capacity, and they have a chip structure that maintains the information as long as power is maintained. They are not large enough to be used for the main memory of a server.

2.2.2. DRAMs

DRAMs (Dynamic RAMs) are the only choice for servers. The term "dynamic" indicates that the information is stored on capacitors within an integrated circuit. Since capacitors discharge over time, due to leakage currents, the capacitors need to be recharged ("refreshed") periodically to avoid data loss. The memory controller is normally in charge of the refresh operations.

2.2.3. SDRAMs

SDRAMs (Synchronous DRAMs) are the most commonly used DRAM. SDRAMs have a synchronous interface, meaning that their operation is synchronized with a clock signal. The clock is used to drive an internal finite state machine that pipelines memory accesses. Pipelining means that the chip can accept a new memory access before it has finished processing the previous one. This greatly improves the performance of SDRAMs compared to classical DRAMs.

DDR2 and DDR3 are the two most commonly used SDRAMs (see Section 2.2.8.) [23].

Figure 24 shows the internal architecture of a DRAM chip.

The memory array is composed of memory cells organized in a matrix. Each cell has a row and a column address. Each bit is stored in a capacitor (i.e., storage Element).

To improve performance and to reduce power consumption, the memory array is split into multiple "banks". Figure 25 shows a 4-bank and an 8-bank organization.

DDR2 chips have 4 internal memory banks and DDR3 chips have 8 internal memory banks.

2.2.4. DIMMs

Multiple memory chips need to be assembled together to build a memory subsystem. They are organized in small boards known as DIMMs (Dual Inline

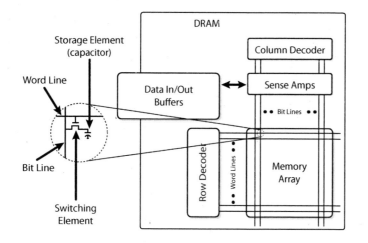

Figure 24: Internal architecture of a DRAM chip

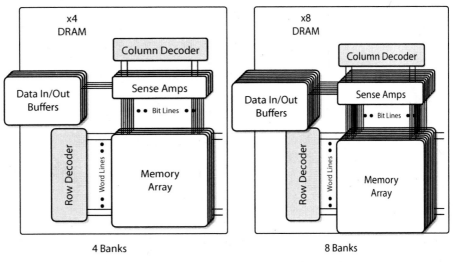

Figure 25: Memory Banks

Memory Modules).

Figure 26 shows the classical organization of a memory subsystem [24]. For example, a memory controller connects four DIMMs each composed of multiple DRAM chips. The memory controller (that may also integrate the clock driver) has an address bus, a data bus, and a command (aka control) bus. It is in charge of reading, writing, and refreshing the information stored in the DIMMs.

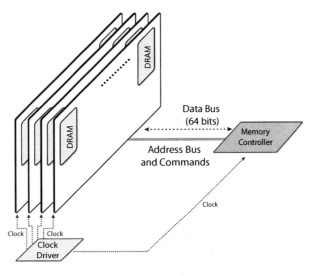

Figure 26: Example of a memory subsystem

Figure 27 is an example of the connection between a memory controller and a DDR3 DIMM. The DIMM is composed of eight DRAM chips, each capable of storing eight bits of data for a total of 64 bits per memory word (width of the memory data bus). The address bus has 15 bits and it carries, at different times, the "row address" or the "column address" for a total of 30 address bits. In addition, three bits of bank address allow accessing the eight banks inside each DDR3 chip. They can be considered equivalent to address bits raising the total addressing capability of the controller to eight Giga words (i.e., 512 Gbits, or 64 GB). Even if the memory controller has this addressing capability, the DDR3 chips available on the market are significantly smaller. Finally RAS (Row Address Selection), CAS (Column Address Selection), WE (Write Enabled), etc. are the command bus wires.

Figure 28 shows a schematic depiction of a DIMM.

The front view shows the eight DDR3 chips each providing four bits of information (normally indicated by "x4"). The side view shows that the chips are on both sides of the board for a total of sixteen chips. i.e., 64 bits.

2.2.5. ECC and Chipkill®

Since data integrity is a major concern in server architecture, very often extra memory chips are installed on the DIMM to detect and recover memory errors. The most common arrangement is to add 8 bits of ECC (Error Cor-

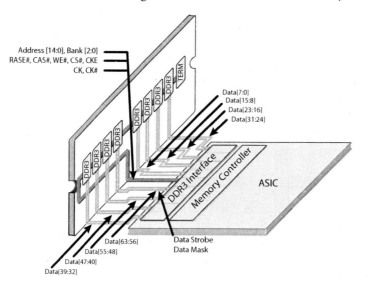

Figure 27: Example of a DDR3 memory controller

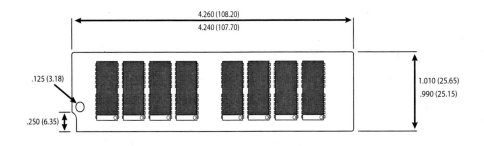

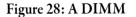

Figure 28: A DIMM

recting Code) to expand the memory word from 64 to 72 bits. This allows the implementation of codes like the Hamming code that allows a single-bit error to be corrected and double-bit errors to be detected. These codes are also known as SEC/DED (Single Error Correction / Double Error Detection).

With a careful organization of how the memory words are written in the memory chips, ECC can be used to protect from any single memory chip that fails and any number of multi-bit errors from any portion of a single memory chip. This feature has several different names [24], [25], [26]:

- Chipkill® is the IBM® trademark
- Sun® Microsystems calls it Extended ECC
- HP® calls it Chipspare®
- A similar system from Intel® is called SDDC® (Single Device Data Correction) or Lockstep Channel Mode.

Chipkill® performs this function by bit-scattering the bits of an ECC word across multiple memory chips, such that the failure of any single memory chip will affect only one ECC bit. This allows memory contents to be reconstructed despite the complete failure of one chip.

While a complete discussion of this technology is beyond the scope of this book an example can give an idea of how it works. Figure 29 shows a memory controller that reads and writes 128 bits of useful data at each memory access, 144 bits when ECC is added. The 144 bits can be divided in 4 memory words of 36 bits and each memory word will be SEC/DED. By using two DIMMs, each with 18 4-bit chips, it is possible to reshuffle the bits as shown in Figure 29.

Now if a chip fails, there will be one error in each of the four words, but since

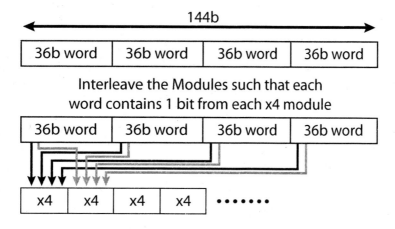

Figure 29: A Chipkill® example

the words are SEC-DEC, each of the four words can correct an error and therefore all the four errors will be corrected.

2.2.6. Memory Ranks

Going back to how the DIMMs are organized, an arrangement of chips that produce 64 bits of useful data (not counting the ECC) is called a "rank". To store more data on a DIMM multiple ranks can be installed. There are single, dual and quad-ranks DIMMs. Figure 30 shows three possible organizations.

In the first drawing a rank of ECC RAM is built using nine eight-bit chips, a configuration that is also indicated 1Rx8. The second drawing shows a 1Rx4 arrangement in which 18 four-bit chips are used to build 1 rank. Finally the third drawing shows a 2Rx8 in which 18 eight-bit chips are used to build 2 ranks.

Memory Ranks are not selected using address bits, but "chip selects". Modern memory controllers have up to eight separate chip selects and therefore are capable of supporting up to eight ranks.

2.2.7. UDIMMs & RDIMMs

SDRAM DIMMs are further subdivided into UDIMMs (Unbuffered DIMMs) and RDIMM (Registered DIMMs). In UDIMMs the memory

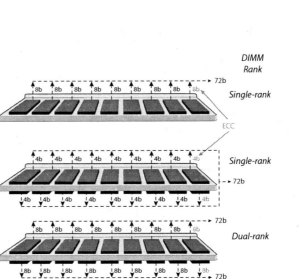

Figure 30: DIMMs and memory ranks

chips are directly connected to the address and control buses, without any intermediate component.

RDIMM have additional components (registers) placed between the incoming address and control buses and the SDRAM components. These registers add one clock cycle of delay but they reduce the electrical load on the memory controller and allow more DIMM to be installed per memory controller.

RDIMM are typically more expensive because of the additional components, and they are usually found in servers where the need for scalability and stability outweighs the need for a low price.

Although any combination of Registered/Unbuffered and ECC/non-ECC is theoretically possible, most server-grade memory modules are both ECC and registered.

Figure 31 shows an ECC RDIMM. The registers are the chips indicated by the arrows; the nine memory chips indicate the presence of ECC.

2.2.8. DDR2 & DDR3

The first SDRAM technology was called SDR (Single Data Rate) to indicate that a single unit of data is transferred per each clock cycle. It was followed by the DDR (Double Data Rate) standard that achieves nearly twice the bandwidth of SDR by transferring data on the rising and falling edges of the clock signal, without increasing the clock frequency. DDR evolved into the two cur-

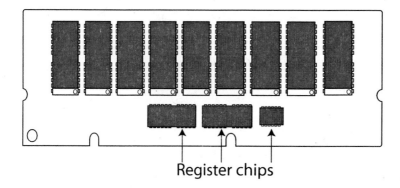

Register chips

Figure 31: ECC RDIMM

rently used standards: DDR2 and DDR3.

DDR2 SDRAMs (double-data-rate two synchronous dynamic random access memories) operate at 1.8 Volts and are packaged in 240 pins DIMM modules. They are capable of operating the external data bus at twice the data rate of DDR by improved bus signaling.

The rules are:

- Two data transfers per DRAM Clock
- Eight bytes (64 bits) per data transfer

Table 1 shows the DDR2 standards[2].

Standard name	DRAM clock	Million data transfers per second	Module name	Peak transfer rate GB/s
DDR2-400	200 MHz	400	PC2-3200	3.200
DDR2-533	266 MHz	533	PC2-4200	4.266
DDR2-667	333 MHz	667	PC2-5300 PC2-5400	5.333
DDR2-800	400 MHz	800	PC2-6400	6.400
DDR2-1066	533 MHz	1,066	PC2-8500 PC2-8600	8.533

Table 1: DDR2 DIMMs

---

2   Some of the DDR2 modules have two names, depending on the manufacturer.

DDR3 SDRAMs (double-data-rate three synchronous dynamic random access memories) improve over DDR2 in the following areas:

- Reduced power consumption obtained by reducing the operating voltage to 1.5 volts;

- Increased memory density by introducing support for chips from 0.5 to 8 Gigabits, i.e. rank capacity up to 16 GB;

- Increased memory bandwidth by supporting a Burst Length = 8, compared to the Burst Length = 4 of DDR2. The reason for the increase in burst length is to better match the increased external data transfer rate with the relatively constant internal access time. As the transfer rate increases, the burst length (the size of the transfer) must increase to not exceed the access rate of the DRAM core.

DDR3 DIMMs have 240 pins, the same number as DDR2, and are the same size, but they are electrically incompatible and have a different key notch location. In the future, DDR3 will also operate at faster clock rate. At the time of publishing, only DDR3-800, 1066, and 1333 are in production.

Table 2 summarizes the different DDR3 DIMM modules.

Standard name	RAM clock	Million data transfers per second	Module name	Peak transfer rate GB/s
DDR3-800	400 MHz	800	PC3-6400	6.400
DDR3-1066	533 MHz	1,066	PC3-8500	8.533
DDR3-1333	667 MHz	1,333	PC3-10600	10.667
DDR3-1600	800 MHz	1,600	PC3-12800	12.800
DDR3-1866	933 MHz	1,866	PC3-14900	14.900

Table 2: DDR3 DIMMs

During 2009 the cost of DDR3 is expected to reach parity with the cost of DDR2.

2.3.   The I/O subsystem

The I/O subsystem is in charge of moving data from the server memory to the external world and vice versa. Historically this has been accomplished by providing in the server motherboards I/O buses compatible with the PCI (Peripheral Component Interconnect) standard. PCI was developed to interconnect peripheral devices to a computer system, it has been around for many years [1] and its current incarnation is called PCI-Express.

2.3.1. PCI-Express

PCI-Express (PCI-E or PCIe) [2], is a computer expansion card interface format designed to replace PCI, PCI-X and AGP.

It removes one of the limitations that have plagued all the I/O consolidation attempts, i.e., the lack of I/O bandwidth in the server buses. It is supported by all current operating systems.

The previous bus-based topology of PCI and PCI-X is replaced by point-to-point connectivity. The resultant topology is a tree structure with a single root complex (in most cases, the CPU itself). The root complex is responsible for system configuration, enumeration of PCIe resources, and manages interrupts and errors for the PCIe tree. A root complex and its endpoints share a single address space and communicate through memory reads and writes, and interrupts.

PCIe connects two components with a point-to-point link. Links are composed of N lanes (a by-N link is composed of N lanes). Each lane contains two pairs of wires: one pair for transmission and one pair for reception.

Multiple PCIe lanes are normally provided by the SouthBridge (aka ICH: I/O Controller Hub) that implements the functionality of "root complex".

Each lane connects to a PCI Express endpoint, to a PCI Express Switch or to a PCIe to PCI Bridge, as in Figure 32.

Different connectors are used according to the number of lanes. Figure 33 shows four different connectors and indicates the speeds achievable with PCIe 1.1

In PCIe 1.1 the lanes run at 2.5 Gbps (2 Gbps at the datalink) and 16 lanes can be deployed in parallel (see Figure 34). This supports speeds from 2 Gbps (1x) to 32 Gbps (16x). Due to protocol overhead 8x is required to support a 10 GE interface.

PCIe 2.0 (aka PCIe Gen 2) doubles the bandwidth per lane from 2 Gbit/s to 4 Gbit/s and extends the maximum number of lanes to 32x. It is shipping at the

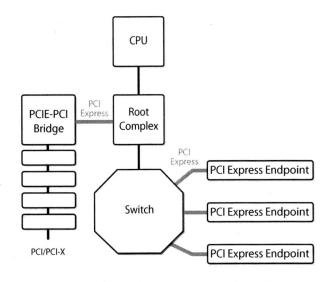

Figure 32: PCI-Express Root Complex

time of writing. 4x is sufficient to support 10 GE.

PCIe 3.0 will approximately double the bandwidth again. The final PCIe 3.0 specifications, including form factor specification updates, may be available by late 2009, and could be seen in products starting in 2010 and beyond [3]. PCIe 3.0 will be required to effectively support 40 GE (Gigabit Ethernet), the next step in the evolution of Ethernet.

All the current deployments of PCI Express are Single Root (SR), i.e., a single

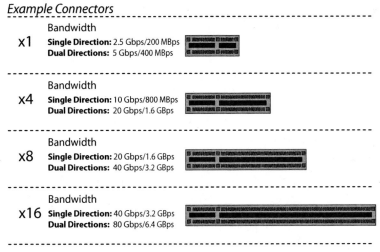

Figure 33: PCI Express Connectors

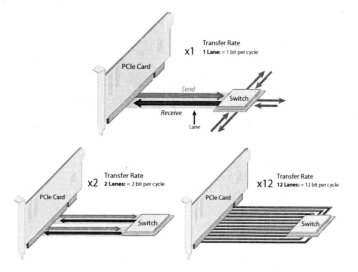

Figure 34: PCI Express lanes

processor controlling multiple endpoints.

Multi Root (MR) has been under development for a while, but it has not seen the light yet, and many question if it ever will. For a discussion of SR in conjunction with virtualization, see Section 3.1.5.

2.4.   Intel® Microarchitecture (Nehalem)

The Intel® Core® i7 Processor is the first of a new family of products implemented using the new 45 nm (nanometer) silicon technology developed by Intel® [32], [33], [34]. First produced in late 2008, these processors (originally known under the codename Nehalem) span the range from high-end desktop applications, up through very large-scale server platforms. Subsequent versions of the processor are being released in 2009 and later. The codename is derived from the Nehalem River on the Pacific coast of northwest Oregon in the United States.

In Intel® parlance processor developments are divided into "tick" and "tock'" intervals, as in Figure 35. Tick is a technology that shrinks an existing processor while tock is a new architecture done in the same technology. Nehalem is the 45nm tock and the 32 nm Westmere tick will follow it.

Nehalem is a balance between different requirements:

- Performance of existing applications compared to emerging appli-

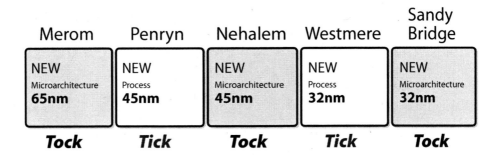

Figure 35: Intel® "Tick/Tock" processor development model

cation (e.g. multimedia);

- Equally good support for applications that are lightly or heavily threaded;

- Implementations that range from laptop to servers.

Nehalem tries to optimize performance and at the same time reduce power consumption. This discussion is based on a nice Intel® Development Forum Tutorial [32].

2.4.1. Platform Architecture

The use of the Intel® QuickPath Interconnect (see Section 2.1.7.) as the foundation of the platform architecture for these processors is the biggest platform architecture shift in about 10 years for Intel. The inclusion of multiple high speed point-to-point connections, along with the use of integrated memory controllers, is a fundamental departure from the FSB-based approach.

An example of a dual-socket Intel Xeon 5500 (Nehalem-EP) systems is shown in Figure 36.

Integrated Memory Controller (IMC)

Each Integrated Memory Controller (IMC) supports three channels of DDR3 (see Section 2.2.8.). DDR3 memories run at higher frequency when compared with DDR2, thus higher memory bandwidth. In addition, for a dual-socket architecture there are two sets of memory controllers instead of one. All these improvements lead to a 3.4x bandwidth increase compared to the previous In-

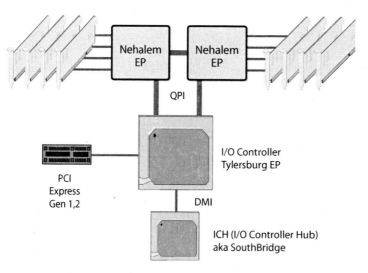

Figure 36: Two-socket Intel Xeon 5500 (Nehalem-EP)

tel® platform (see Figure 37).

This will continue to increase over time, as faster DDR3 becomes available. An integrated memory controller also makes a positive impact by reducing latency.

Power consumption is also reduced, since DDR3 is 1.5 Volt technology compared to 1.8 Volts of DDR2. Power consumption tends to go with the square

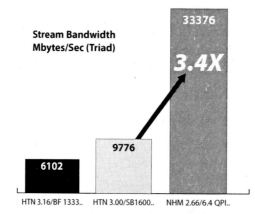

Figure 37: RAM bandwidth

Source: Intel Internal measurements – August 2008
HTN: Intel® Xeon® processor 5400 Series (Harpertown)
NHM: Intel® Core™ microarchitecture (Nehalem)

of the voltage and therefore a 20% reduction in voltage causes approximately a 40% reduction in power.

Finally the IMC supports both RDIMM and UDIMM with single, dual or quad ranks (quad ranks is only supported on RDIMM, see Section 2.2.6. and Section 2.2.7.).

Intel® QuickPath Interconnect (QPI)

All the communication architectures have evolved over time from busses to point-to-point links that are much faster and more scalable. In Nehalem, Intel® QuickPath Interconnect has replaced the Front Side Bus (see Figure 38).

Intel® QuickPath Interconnect is a coherent point-to-point protocol introduced by Intel®, not limited to any specific processor, to provide communication between processors, I/O devices and potentially other devices such as accelerators.

2.4.2. CPU Architecture

Nehalem increases the instructions per second of each CPU by a number of innovations depicted in Figure 39.

Some of these innovations are self-explanatory; we will focus here on the most

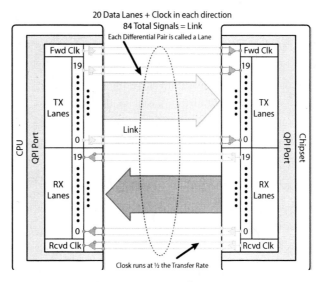

Figure 38: Intel® QPI

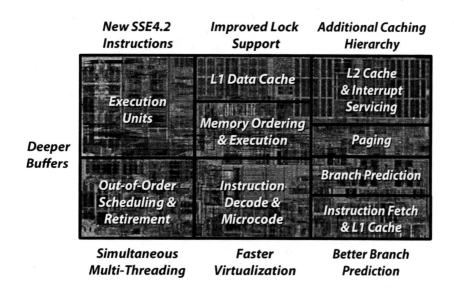

Figure 39: Nehalem microarchitecture innovations

important one that deals with performance vs. power.

In comparing performance and power it is normally assumed that a 1% performance increase with a 3% power increase is break-even. The reason is that it is always possible to reduce the voltage by 1% and reduce the power by 3% (see Section 2.4.4.).

The innovations that are important are those that improve the performance by 1% with only a 1% increase in power (better than break-even).

Intel® Hyper-Threading Technology (Intel® HT Technology)

Another innovation is Intel® Hyper-Threading Technology (Intel® HT Technology), i.e. simultaneous multi-threading which enhances both performance and energy efficiency (see Section 2.1.4.).

The basic idea is that with the growing complexity of each execution unit, it is difficult for a single thread to keep the execution unit busy. Instead by overlaying two threads on the same core it is more likely that all the resources can be kept busy and therefore the overall efficiency increases (see Figure 40). Hyper-Threading consumes a very limited amount of area (less than 5%) and it is extremely effective in increasing efficiency, in a heavily threaded environment. Hyper-Threading is not a replacement for cores; it complements cores by allowing each of them to execute two threads simultaneously.

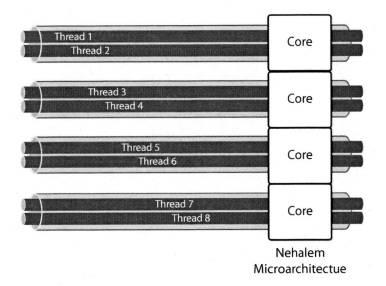

Nehalem
Microarchitectue

Figure 40: Intel® HT Technology

Cache-Hierarchy

The requirement of an ideal memory system is that it should have infinite capacity, infinite bandwidth, and zero latency. Of course, nobody knows how to build such a system. The best approximation is a hierarchy of memory subsystems that go from larger and slower to smaller and faster. In Nehalem Intel® added one level of hierarchy by increasing the cache layers from two to three, see Figure 41.

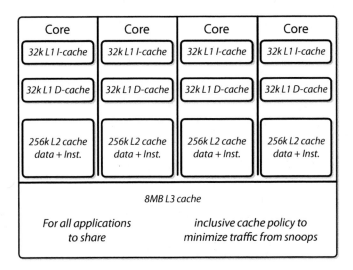

Figure 41: Cache Hierarchy

Level one caches (L1) (Instruction and Data) are basically unchanged compared to previous Intel® designs. In the previous Intel® design the level two caches (L2) were shared across the cores. This was possible since the number of cores was limited to two. Nehalem increments the number of cores to four or eight and the L2 caches cannot be shared any longer, due to the increase in bandwidth and arbitration requests (potentially 8X). For this reason in Nehalem Intel® added L2 caches (Instruction and Data) dedicated to each core to reduce the bandwidth toward the shared caches that is now a level three (L3) cache.

Segmentation

Nehalem is designed for modularity. Cores, caches, IMC, Intel® QPI are examples of modules that compose a Nehalem processor (see Figure 39).

These modules are designed independently and they can run at different frequencies and different voltages. The technology that glues all of them together is a novel synchronous communication protocol that provides very low latency. Previous attempts used asynchronous protocols that are less efficient.

Integrated Power Gate

This is a power management technique that is an evolution of the "Clock Gate" that exists in all modern Intel® processors. The clock gate shuts off the

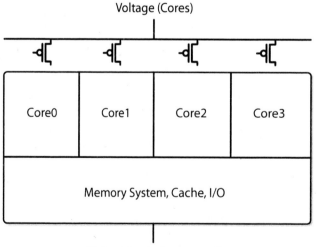

Figure 42: Nehalem Power Gate

clock signals to idle logic, thus eliminating switching power, but leakage current remains. Leakage currents create leakage power consumption that has no useful purpose. With the reduction in channel length, starting approximately at 130 nm, leakage has become a significant part of the power and at 45 nm it is very important.

The power gate instead shuts off both switching and leakage power and enables an idle core to go to almost zero power (see Figure 42). This is completely transparent to software and applications.

The power gate is difficult to implement from a technology point of view. The classical elements of 45 nm technologies have significant leakage. It required a new transistor technology with a massive copper layer (7mm) that was not done before (see Figure 43).

The power gate becomes more important as the channel length continues to shrink, since the leakage currents continue to increase. At 22 nm, the power gate is essential.

Power Management

Power sensors are key in building a power management system. Previous Intel® CPUs had thermal sensors, but they didn't have power sensors. Nehalem has both thermal and power sensors that are monitored by an integrated micro-controller (PCU) in charge of power management (see Figure 44).

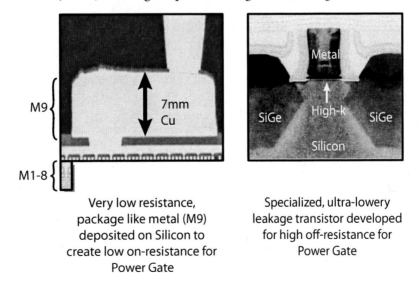

Very low resistance, package like metal (M9) deposited on Silicon to create low on-resistance for Power Gate

Specialized, ultra-lowery leakage transistor developed for high off-resistance for Power Gate

Figure 43: Power Gate transistor

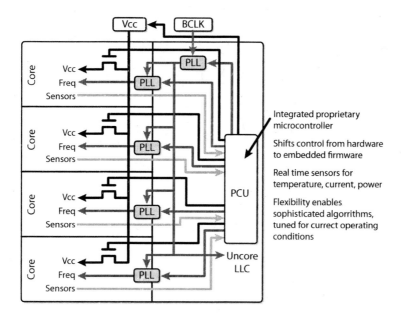

Figure 44: Power Control Unit

Intel® TurboBoost Technology

Power gates and Power management are the basic components of Intel® TurboBoost Technology. TurboBoost mode is used when the operating system requires more performance, if environmental conditions permit (sufficient cooling and power), for example because one or more cores are turned off. TurboBoost increases the frequency of the active cores (and also the power consumption), thus increasing the performance of a given core (see Figure 45). This is not a huge improvement (from 3% to 11%), but it may be particularly valuable in lightly or non-threaded environments in which not all the cores may be used in parallel. The frequency is increased in 133MHz steps.

Figure 45 shows three different possibilities: in the normal case all the cores operate at the nominal frequency (2.66 GHz), in the "4C Turbo" mode all the cores are frequency upgraded by one step (for example to 2.79 GHz), and in the "<4C Turbo" mode two cores are frequency upgraded by two steps (for example to 2.93 GHz).

2.4.3. Virtualization support

Intel® Virtualization Technology (VT) extends the core platform architecture to better support virtualization software, e.g., VMs (Virtual Machines) and

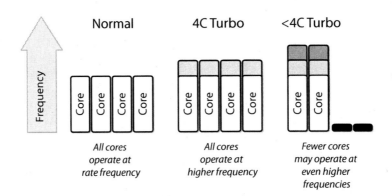

Figure 45: Intel® TurboBoost Technology

hypervisors aka VMMs (Virtual Machine Monitors), see Figure 46.

VT has three major components:

- Intel® VT-x refers to all the hardware assists for virtualization in Intel® 64 and IA32 processors;

- Intel® VT for Directed I/O (Intel® VT-d) refers to all the hardware assists for virtualization in Intel® chipset;

- Intel® VT for Connectivity (Intel® VT-c) refers to all the hardware assists for virtualization in Intel® networking and I/O devices.

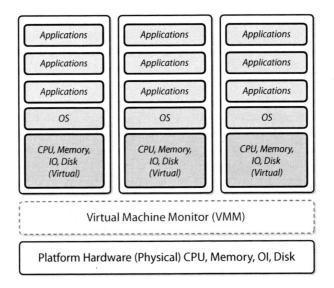

Figure 46: Virtualization Support

Intel® VT-x enhancements include:

- A new, higher privilege ring for the hypervisor — This allows guest operating systems and applications to run in the rings they were designed for, while ensuring the hypervisor has privileged control over platform resources.

- Hardware-based transitions — Handoff between the hypervisor and guest operating systems are supported in hardware. This reduces the need for complex, compute-intensive software transitions.

- Hardware-based memory protection — Processor state information is retained for the hypervisor and for each guest OS in dedicated address spaces. This helps to accelerate transitions and ensure the integrity of the process.

In addition Nehalem adds:

- EPT (Extended Page Table);
- VPID (Virtual Processor ID);
- Guest Preemption Timer;
- Descriptor Table Exiting;
- FlexMigration.

Extended Page Tables (EPT)

EPT is a new page-table structure, under the control of the hypervisor (see Figure 47). It defines mapping between guest- and host-physical addresses.

Before virtualization, each OS was in charge of programming page tables to translate between virtual application addresses and the "physical addresses".

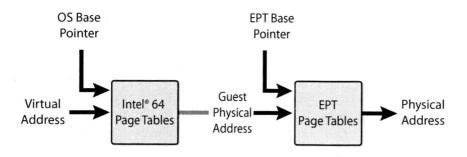

Figure 47: Extended Page Tables

With the advent of virtualization these addresses are no longer physical, but instead local to the VM. The hypervisor needs to translate between the guest OS addresses and the real physical addresses. Before EPT, the hypervisors maintained the page table in software by updating them at significant boundaries (e.g. on VM entry and exit).

With EPT, there is an EPT base pointer and an EPT Page Table that allow it to go directly from the virtual address to the physical address without the hypervisor intervention, in a way similar to how an OS does it in a native environment.

Virtual Processor ID (VPID)

This is the ability to assign a VM ID to tag CPU hardware structures (e.g. TLBs: Translation Lookaside Buffers) to avoid flushes on VM transitions.

Before VPID, in a virtualized environment, the CPU flushes the TLB unconditionally for each VM transition (e.g. VM Entry/Exit). This is not efficient and adversely affects the CPU performance. With VPID, TLBs are tagged with an ID decided by the hypervisor that allows a more efficient flushing of cached information (only flush what is needed).

Guest Preemption Timer

A mechanism for a hypervisor to preempt execution of a guest OS.

With this feature, a hypervisor can preempt guest execution after a specified amount of time. The hypervisor sets a timer value before entering a guest and when the timer reaches zero, a VM exit occurs. The timer causes a VM exit directly with no interrupt, so this feature can be used with no impact on how the VMM virtualizes interrupts.

Descriptor Table Exiting

Allows a VMM to protect a guest OS from internal attack by preventing relocation of key system data structures.

OS operation is controlled by a set of key data structures used by the CPU: IDT , GDT LDT, and TSS. Without this feature, there is no way for the hypervisor to prevent malicious software running inside a guest OS from modifying the guest's copies of these data structures. A hypervisor using this feature can intercept attempts to relocate these data structures and forbid malicious ones.

FlexMigration

FlexMigration allows migration of the VM between processors that have a different instruction set. It does that by synchronizing the minimum level of the instruction set supported by all the processors in a pool.

When a VM is first instantiated it queries its processor to obtain the instruction set level (SSE2, SSE3, SSE4). The processor returns the agreed minimum instruction set level in the pool, not the one of the processor itself.

2.4.4. Chip Design

When trying to achieve high performance and limit the power consumption several different factors need to be balanced.

With the progressive reduction of the length of the transistor channel the range of voltages usable becomes limited (see Figure 48).

The maximum voltage is limited by the total power consumption and the reliability decrease associated with high power, the minimum voltage is limited mostly by soft errors especially in memory circuits.

In general, in CMOS design the performance is proportional to the voltage, since higher voltages allow higher frequency.

Performance ~ Frequency ~ Voltage

Power consumption is proportional to the frequency and the square of voltage:

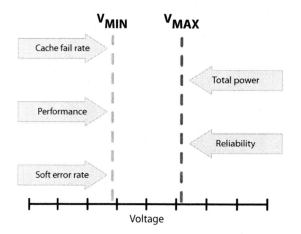

Figure 48: Voltage Range

$$Power \sim Frequency \ x \ Voltage^2$$

and, since Frequency and Voltage are proportional:

$$Power \sim Voltage^3$$

Energy efficiency is defined as the ratio between performance and power and therefore:

$$Energy \ Efficiency \sim 1/Voltage^2$$

Therefore from an energy efficiency perspective there is an advantage in reducing the Voltage (i.e., the power, see Figure 49) so big that Intel® has decided to address it.

Since the circuits that are more subject to soft error are memories, Intel® in Nehalem deploys a sophisticated error correcting code (triple detect, double correct) to compensate for these soft errors. Also the voltage of the caches and the voltage of the cores are decoupled so the cache can stay at high voltage while the cores works at low voltage.

For the L1 and L2 caches Intel® has replaced the traditional six transistors SRAM design (6-T SRAM) with a new eight transistors design (8-T SRAM) that decouples the read and write operations and allows lower voltages (see Figure 50).

Also to reduce power Intel® went back to static CMOS, which is the CMOS technology that consumes less power (see Figure 51).

Performance was regained by redesigning some of the key algorithm like instruction decoding.

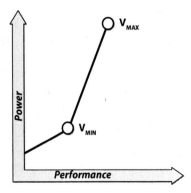

Figure 49: Power vs. Performance

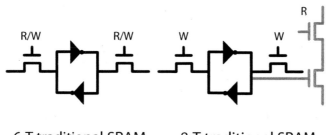

6-T traditional SRAM   8-T traditional SRAM
High Voltage   Low Voltage friendly

Figure 50: Six- vs. eight-transistor SRAM

2.5.   Virtualization Support

In addition to the virtualization support provided inside Nehalem, other improvements have been implemented at the chipset/motherboard level to better support virtualization. These improvements are important to increase the I/O performance in the presence of a hypervisor (in Intel® parlance the hypervisor is referred to as VMM: Virtual Machine Monitor).

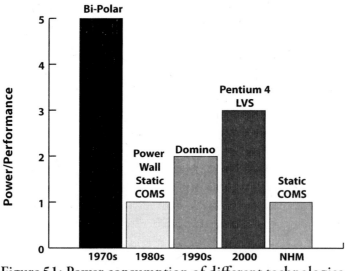

Figure 51: Power consumption of different technologies

2.5.1. *VMDq*

In virtual environments, the hypervisor manages network I/O activities for all the VMs (Virtual Machines). With the constant increase in the number of VMs the I/O load increases and the hypervisor requires more CPU cycles to sort data packets in network interface queues and route them to the correct VM, reducing CPU capacity available for applications.

Intel® Virtual Machine Device Queues (VMDq) reduces the burden on the hypervisor while improving network I/O by adding hardware support in the chipset. In particular multiple network interface queues and sorting intelligence are added to the silicon as shown in Figure 52.

As data packets arrive at the network adapter, a Layer 2 classifier/sorter in the network controller sorts and determines which VM each packet is destined for based on MAC addresses and VLAN tags. It then places the packet in a receive queue assigned to that VM. The hypervisor's layer 2 software switch merely routes the packets to the respective VM instead of performing the heavy lifting work of sorting data.

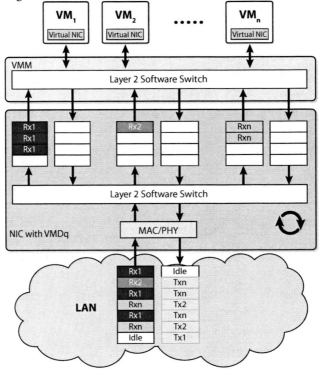

Figure 52: VMDq

As packets are transmitted from the virtual machines towards the adapters, the hypervisor layer places the transmit data packets in their respective queues. To prevent head-of-line blocking and ensure each queue is fairly serviced, the network controller transmits queued packets to the wire in a round-robin fashion, thereby guaranteeing some measure of Quality of Service (QoS) to the VMs.

2.5.2. NetQueue

To take full advantage of VMDq the VMMs needs to be modified to support one queue per Virtual Machine. For example, VMware® has introduced in its hypervisor a feature called NetQueue that takes advantage of the frame sorting capability of VMDq. The combination of NetQueue and VMDq offloads the work that ESX has to do to route packets to virtual machines, therefore it frees up CPU and reduces latency.

2.5.3. VMDirectPath

When VMDq is combined with VMware® VMDirectPath, the hypervisor is bypassed in a way similar to a "kernel bypass", the software switch inside the hypervisor is not used, and data is directly communicated from the adapter to the vNICs (virtual NICs) and vice versa (see Figure 53).

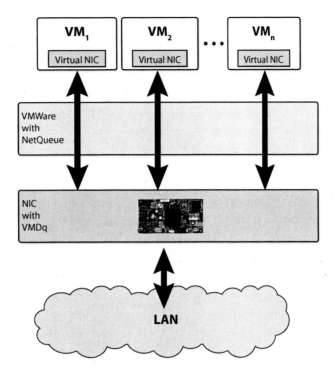

Figure 53: VMM NetQueue

3. California enabling technologies

3.1. Unified Fabric

The terms Unified Fabric or I/O consolidation are synonymous and refer to the ability of a network (both switches and host adapters) to use the same physical infrastructure to carry different types of traffic that typically have different traffic characteristics and handling requirements.

From the network side this equates in having to install and operate a single network instead of three. From the hosts side, fewer CNAs (Converged Network Adapters) replace and consolidate NICs (Network Interface Cards), HBAs (Host Bus Adapters) and HCAs (Host Channel Adapters). This results in a lower number of PCIe slots required on rack mounted servers and it is particularly beneficial in the case of blade servers where often only a single mezzanine card is supported per blade.

Customers benefits are:

- great reduction, simplification and standardization of cabling;
- absence of gateways that cause bottleneck and are a source of incompatibilities;
- less power and cooling;
- reduced cost.

Figure 54 shows an example where 2 FC HBAs, 3 Ethernet NICs and 2 IB HCAs are replaced by 2 CNAs (Converged Network Adapters).

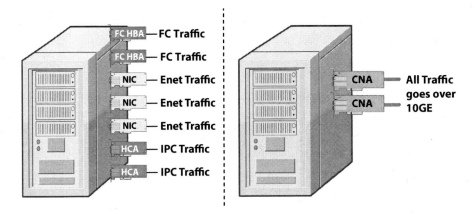

Figure 54: Server with Unified Fabric connections

The biggest challenge of I/O consolidation is to satisfy the requirements of different traffic classes within a single network without creating "traffic interference", i.e., the possibility of one class of traffic to stave another.

Deployments worldwide of IPv4 and IPv6 on Ethernet based networks has resulted in it becoming the de facto standard for all LAN traffic [4]. Too much investment has been done in this area and too many applications assume that Ethernet is the underlying network, for this to change. This traffic is characterized by a large number of flows. Typically these flows are not very sensitive to latency, but this is changing rapidly and latency must be taken into serious consideration. Streaming Traffic is also sensitive to latency jitter.

Storage traffic must follow the Fibre Channel (FC) model. Again, large customers have massive investment in FC infrastructure and management. Storage provisioning often relies on FC services like naming, zoning, etc. In FC losing frames is not an option, since SCSI is extremely sensitive to packet drops. This traffic is characterized by large packet sizes, typically 2KB in payload.

IPC (Inter Processor Communication) traffic is characterized by a mix of large and small messages. It is typically latency sensitive, especially the short messages. IPC traffic is used in "Clusters", i.e., interconnections of two or more computers. In the data center, examples of server clustering include:

- Availability clusters (e.g., Symantec/Veritas VCS, MSCS);

- Clustered File Systems;

- Clustered Databases (e.g., Oracle RAC);

- VMware® Virtual Infrastructure Services (e.g., VMware® VMotion, VMware® HA).

Cluster technologies usually require a separate network fabric to guarantee that normal network traffic does not interfere with latency requirements. Clusters are not as dependent on the underlying network technology, provided that it is low cost, it offers high bandwidth, it is low latency and the adapters provide zero-copy mechanisms (they avoid making intermediates copies of packets).

3.1.1. 10 Gigabit Ethernet

10GE (10 Gigabit Ethernet) is the network of choice for I/O consolidation. During 2008 the standard reached the maturity status and low cost cabling solutions are available. Fiber continues to be used for longer distances, but copper is deployed in the Data Center to reduce the costs.

Switches and CNAs have standardized their connectivity using the small form factor plus pluggable (SFP+) transceiver. SFP+ are used to interface a network device mother board (switches, routers or CNAs) to a fiber optic or copper cable.

Unfortunately the IEEE standard for Twisted Pair (10GBASE-T) requires an enormous number of transistors, especially when the distance approaches 100 meters (328 feet). This translates to a significant power requirement and also into additional delay (Figure 55). Imagine trying to cool a line card that has 48 10GBASE-T ports on the front-panel, each consuming 4 Watts!

A more practical solution in the Data Center, at the rack level, is to use SFP+ with Copper Twinax cable (defined in Annex of SFF-8431, see [6]). This cable is very flexible; approximately 6 mm (1/4 of an inch) in diameter and it uses the SFP+ as the connectors. Cost is limited, power consumption and delay are negligible. Copper Twinax cables are limited to 10 meters (33 feet) that is sufficient to connect a few racks of servers to a common top of the rack switch.

3.1.2. Lossless Ethernet

To satisfy the requirements of Unified Fabric, Ethernet has been enhanced to

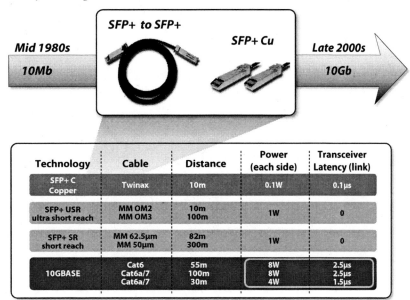

Technology	Cable	Distance	Power (each side)	Transceiver Latency (link)
SFP+ C Copper	Twinax	10m	0.1W	0.1µs
SFP+ USR ultra short reach	MM OM2 MM OM3	10m 100m	1W	0
SFP+ SR short reach	MM 62.5µm MM 50µm	82m 300m	1W	0
10GBASE	Cat6 Cat6a/7 Cat6a/7	55m 100m 30m	8W 8W 4W	2.5µs 2.5µs 1.5µs

Figure 55: 10GE cabling options

operate both as a "lossless" and "lossy" network. This classification does not consider losses due to transmission errors that, in a controlled environment with limited distances like the Data Center, are rare compared to losses due to congestion.

Fibre Channel and IB (InfiniBand) are examples of lossless networks: they have a link level signaling mechanism to keep track of buffer availability at the other end of the link. This mechanism allows the sender to send a frame only if a buffer is available at the receiving end and therefore the receiver never needs to drop frames. Avoiding frame drops is mandatory for carrying native storage traffic over Ethernet, since storage traffic does not tolerate frame loss. SCSI was designed with the assumption that SCSI transactions are expected to succeed and that failures are so rare that is acceptable to recover slowly in such an event.

3.1.3. Terminology

DCB (Data Center Bridging) is a term used by IEEE to collectively indicate the techniques described in Section 3.1.4., Section 3.1.5. , and Section 3.1.6..

The terms CEE (Converged Enhanced Ethernet) and DCE (Data Center Ethernet) have also been used to group these techniques under common umbrellas.

3.1.4. PFC (Priority Flow Control)

Following these requirements Ethernet has been enhanced by IEEE 802.1bb: a physical link can be partitioned into multiple logical links (by extending the IEEE 802.1Q Priority concept) and each priority can be configured to have either a lossless or a lossy behavior. This new mechanism is called PFC (Priority Flow Control), aka PPP (Per Priority Pause) [8].

With PFC, an administrator can define which priorities are lossless and which are lossy. The network devices will treat the lossy priorities as in classical Ethernet and will use a per priority pause mechanism to guarantee that no frames are lost on the lossless priorities.

If separate traffic classes are mapped to different priorities, there is no traffic interference. For example, in Figure 56, storage traffic is mapped to priority three and it is paused, while IPC traffic is mapped to priority six and it is being forwarded and so is IP traffic that is mapped to priority one.

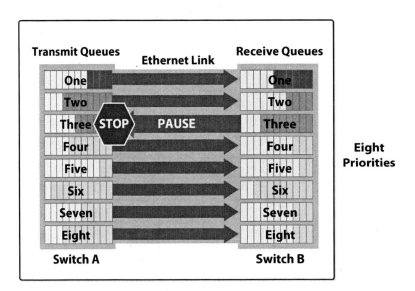

Figure 56: Priority Flow Control

PFC requires a more complex organization in the data plane that allows for resources, such as buffers or queues, to be allocated on a per priority basis.

PFC is the basic technique required to implement I/O consolidation. Additional techniques make I/O consolidation deployable on a larger scale. The next paragraphs describe two additional components:

- Discovery Protocol (DCBX);

- Bandwidth Manager (aka ETS: Enhanced Transmission Selection).

3.1.5. DCBX: Data Center Bridging eXchange

DCBX is a discovery and configuration protocol that guarantees that both ends of an Ethernet link are configured consistently. This is to avoid "soft errors" which can be very difficult to troubleshoot.

DCBX discovers the capabilities of the two peers at each end of a link: it can check for consistency, it can notify the device manager in the case of configuration mismatches, and it can provide basic configuration in the case where one of the two peers is not configured. DCBX can be configured to send conflict alarms to the appropriate management stations.

3.1.6. Bandwidth Management

IEEE 802.1Q-2005 defines eight priorities, but not a simple, effective and consistent scheduling mechanism. The scheduling goals are typically based upon bandwidth, latency and jitter control.

Products typically implement some form of Deficit Weighted Round Robin (DWRR), but there is no consistency across implementations, and therefore configuration and interworking is problematic.

IEEE is standardizing on a hardware efficient two-level DWRR with strict priority support in IEEE 802.1Qaz ETS (Enhanced Transmission Selection) [12].

With this structure, it is possible to assign bandwidth based on traffic classes, for example: 40% LAN, 40% SAN, and 20% IPC. This architecture allows control not only of bandwidth, but also of latency. Latency is becoming increasingly important for IPC applications. An example of link bandwidth allocation is shown in Figure 57.

3.1.7. FCoE (Fibre Channel over Ethernet)

FCoE is described in detail in a book from Silvano Gai and Claudio DeSanti [17].

FCoE (Fibre Channel over Ethernet) is a standard developed by the FC-BB-5

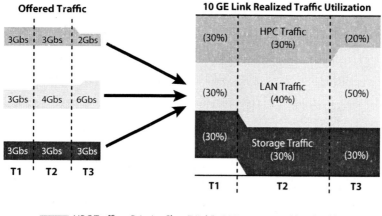

HPC Traffic – Priority Class "High"– 20% guaranteed bandwidth

LAN Traffic – Priority Class "Medium"– 50% guaranteed bandwidth

Storage Traffic – Priority Class "Medium High"– 30% default bandwidth

Figure 57: Bandwidth Manager

working group of INCITS T11 [13], [14], [15]. FCoE, being based on the FC protocol dominant in storage networks, is able to provide a true I/O consolidation solution based on Ethernet.

The idea behind FCoE is simple — to implement I/O consolidation by carrying each FC frame inside an Ethernet frame. The encapsulation is done on a frame-by-frame basis and therefore keeps the FCoE layer completely stateless and it does not require fragmentation and reassembly.

FCoE traffic shares the physical Ethernet links with other traffic, but it is runs on a lossless priority to match the lossless behavior guaranteed in Fibre Channel by buffer-to-buffer credits.

Figure 58 shows an example of I/O consolidation using FCoE in which the only connectivity needed on the server is Ethernet, while separate backbones can still be maintained for LAN and SAN.

FCoE has the advantage of being completely part of the Fibre Channel architecture and it is therefore able to provide seamless integration with existing FC SANs, allowing reuse of existing FC SAN tools and management constructs.

Figure 59 also shows that an FCoE-connected server is just a SCSI initiator over FCP/FC, exactly as if the server were connected over native FC. The same applies to FCoE connected Storage Arrays: they act just as SCSI targets.

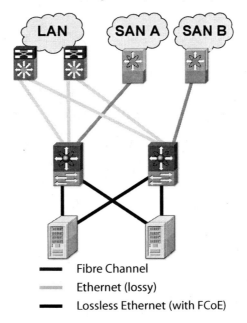

Figure 58: Example of Unified Fabric with FCoE

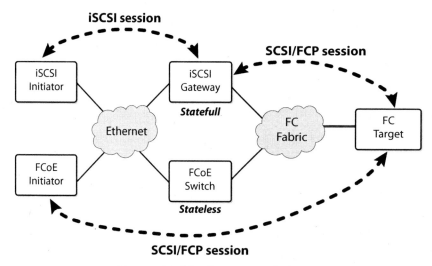

Figure 59: No Gateway in FCoE

Another notable advantage of FCoE is that it requires no gateway. In fact the encapsulation/de-encapsulation functions simply add or remove an Ethernet envelope around a FC frame. The FC frame remains untouched and the operation is completely stateless. Figure 59 shows a comparison of FCoE with iSCSI from the absence/presence of a gateway.

The management tools that customers use to manage and maintain their SANs today can be used in an FCoE environment. From a storage administrator perspective, zoning is a basic provisioning function that is used to give hosts access to storage. FCoE switches provide the same unmodified zoning functionality ensuring that storage allocation and security mechanisms are unaffected. The same consideration applies to all other Fibre Channel services such as dNS, RSCN and FSPF.

The FCoE encapsulation is shown in Figure 60. Starting from the inside out, there is the FC Frame that can be up to 2 KB, hence the requirement to support jumbo frame sizes up to 2.5KB. The FC frame contains the original FC-CRC. This is extremely important, since the FC frame and its CRC remain unmodified end-to-end, whether they are being carried over FC or over FCoE. Next is the FCoE header and trailer that mainly contain the encoded FC start of frame and end of frame (in native FC these are ordered sets that contain code violation and therefore they need to be re-encoded, since a code violation cannot appear in the middle of a frame). Finally the Ethernet header which contains Ethertype = FCoE and the Ethernet trailer that contains the FCS (Frame Control Sequence). The reserved bits have been inserted so that the

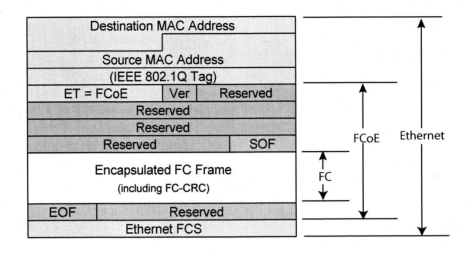

Figure 60: FCoE frame format

Ethernet payload is greater or equal to the minimum Ethernet Payload (46 bytes) in the presence of minimum size FC frames (28 bytes).

In FC all the links are point-to-point, while Ethernet switches creates "clouds". With the term cloud we refer to an Ethernet multi-access network, a broadcast domain (not to be confused with cloud computing).

Figure 61 shows a classical blade server deployment where two Ethernet switches inside the blade server create two "Ethernet Clouds" to which multiple FCoE capable end stations are connected. It is therefore necessary to discover among the end stations connected to a cloud that are FCoE capable. For this reason FCoE is really two different protocols:

- FCoE itself is the data plane protocol. It is used to carry most of the FC frames and all the SCSI traffic. This is data intensive and typically it is switched in hardware.

- FIP (FCoE Initialization Protocol) is the control plane protocol. It is used to discover the FC entities connected to an Ethernet cloud and by the hosts to login to and logout from the FC fabric. This is not a data intensive protocol and it is typically implemented in software on the switch supervisor processor.

The two protocols have two different Ethertypes. Figure 62 shows the steps of FIP that lead to a successful FLOGI and to the possibility of exchanging SCSI traffic using FCoE.

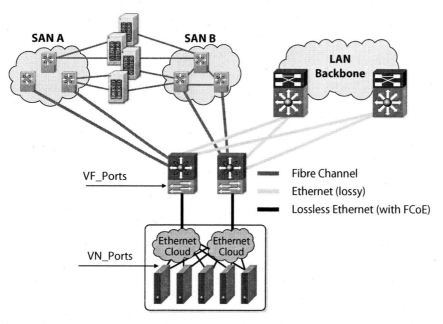

Figure 61: Ethernet Clouds inside blade servers

Once the initialization phase is completed, virtual ports are created. The term "virtual" refers to an FC port that is implemented on a network that is not native FC. Therefore FC-BB-5 defines:

- VN_Port (Virtual N_Port): an N_Port over an Ethernet link;

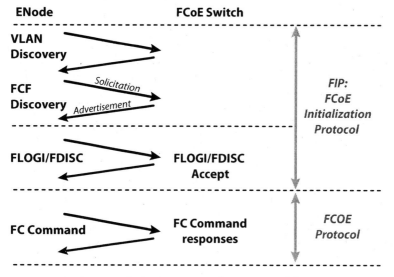

Figure 62: FCoE Initialization Ladder

- VF_Port (Virtual F_Port): an F_Port over an Ethernet link;
- VE_Port (Virtual E_Port): an E_Port over an Ethernet link.

Figure 61 shows the position of the VF_Ports and VN_Port in a classical blade server deployment.

The ENode typically implements the VN_Port. ENodes are commercially called CNAs (Converged Network Adapters). It is a PCI Express adapter that contains both HBA and NIC functionalities, as shown in Figure 63.

The unification occurs only on the network facing side, since two separate PCI-Express devices are still presented to the server. Figure 64 shows how Windows views a dual port CNA within domain manager:

- In the "Network Adapters" section of the Device Manager it sees two Ethernet NICs;
- In the "SCSI and RAID controllers" section of the Device Manager it sees two Fibre Channel Adapters.

From the SCSI perspective, the host is incapable of distinguishing whether SCSI is running over FCoE or native FC.

The high availability of Ethernet will be dealt with at a lower level using NIC teaming or bonding, while the high availability of Fibre Channel will be dealt with at a much higher level using storage multipath software.

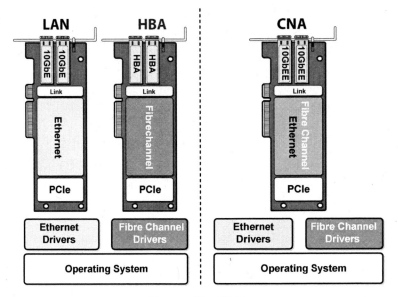

Figure 63: CNAs

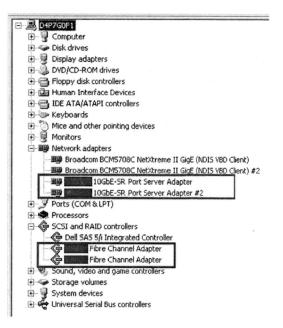

Figure 64: Windows view of a CNA

CNAs are available as regular PCI Express boards or in mezzanine form factor for blade servers. CNAs will soon start to appear on the motherboard in an arrangement often referred to as LOM (LAN On Motherboard). As LOM becomes available there will be an even greater adoption of FC, as businesses leverage the cost benefits that LOM provides.

FCoE can also be implemented entirely in software. This may be particularly interesting for servers that perform non-I/O intensive operations, but that have a need to access storage arrays.

A complete software implementation of FCoE exists in an open-source project (www.Open-FCoE.org) [13]. This will soon start to appear in Linux distributions such as RedHat® or SuSE® and later possibly also in Microsoft® Windows®.

3.2. Virtualization

One of the most overused terms in the computer and network industry is "virtualization". It is used in conjunction with servers, platforms, resources, applications, desktops, networks, etc.

The following sections describe the relation between server virtualization and network virtualization and how this is relevant to the California project.

3.2.1. Server Virtualization

The times when a server had a single CPU, a single Ethernet card (with a unique MAC address) and a single operating system are long gone. Today servers are much more complex. They contain multiple sockets (see Section 2.1.1.) with each socket containing multiple cores (see Section 2.1.2.) and each core being capable of running one or more threads simultaneously. These servers have significant I/O demands and they use multiple NICs (Network Interface Cards) to connect to various different networks, and to guarantee performance and high availability. These NICs are evolving to support SR-IOV (see Section 3.2.2.) and server virtualization.

Server virtualization is a technique that allows use of all the available cores without modifying/rewriting the applications. VMware® ESX®, Linux® XEN® and Microsoft® Hyper-V® are well known virtualization solutions that enable multiple Virtual Machines (VMs) on a single server through the coordination of a hypervisor.

A VM is an instantiation of a logical server that behaves as a standalone server, but it shares the hardware and network resources with the other VMs.

The hypervisor implements VM to VM communication using a "software switch" module, his creates a different model compared to standalone servers. Standalone servers connect to one or more Ethernet switches through dedicated switch ports (see Figure 65). Network policies applied to these Ethernet switch ports (dashed line in Figure 65) are effectively applied to the single

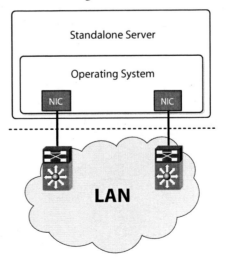

Figure 65: Standalone Server

standalone servers.

A logical server running in a VM connects to the software switch module in the hypervisor and this in turn connects to one or more Ethernet switches (see Figure 66). Network policies applied to the Ethernet switch ports (dashed line in Figure 66) are not very effective, since they are applied to all the VMs (i.e., logical servers) and cannot be differentiated per VM.

Attempts to specify such policies in term of source MAC addresses are also not effective, since the MAC addresses used by VMs are assigned by the virtualization software and can change over time. Moreover, this opens the system to MAC address spoofing attacks, for example, a VM may try to use the MAC address assigned to another VM.

Virtualization software can also move the VMs between different physical servers making the management of network policies even more challenging. The number of VMs tends to be much larger than the number of physical servers and this creates scalability and manageability concerns. For these reasons, alternative solutions have been identified to avoid using a software switch inside the hypervisor, but still allowing VM to VM communication.

3.2.2. SR-IOV

The PCI-SIG (Peripheral Component Interconnect - Special Interest Group) has a subgroup that produces specifications for NIC cards supporting I/O

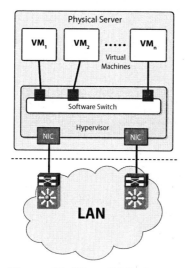

Figure 66: Virtual Machines

Virtualization (IOV). SR-IOV (Single Root IOV) deals with native I/O Virtualization in PCI Express topologies where there is a single root complex. These devices are designed to work in conjunction with server virtualization by allowing multiple VMs to share PCI Express devices, potentially bypassing the hypervisor (Section 2.3.1.).

3.2.3. The IEEE standard effort

IEEE has been in charge of standardizing Ethernet switching behavior with the project IEEE 802.1, that has been recently extended to cover the VEBs (Virtual Ethernet Bridges). To facilitate this, two options are available:

- Move the policy enforcement and the management inside the adapter (see Section 3.2.4.)
- Carry the VM identity to the Ethernet Switch (see Section 3.2.2.).

Cisco Systems has also started to offer a new architecture called VN-Link (see Section 3.2.8.) which contains products that are compliant with this IEEE effort.

3.2.4. VEB in the adapter

The term *"VEB in the adapter"* refers to the implementation of the VEB inside the adapter (NIC) (see Figure 67). This implies that large NIC manufacturers like Intel® and Broadcom® should start to integrate Ethernet switches, capable of enforcing policies, in their NICs.

At a first glance this solution may look attractive, since it does not require changing the frame format on the wire between the NIC and the Ethernet switch, but it has significant drawbacks. The NIC has to implement all the functions of an Ethernet switch, with all the security components. This increases cost and complexity and subtracts valuable gates to other ULP (Upper Layer Protocol) features like TCP offload, RDMA, FC/SCSI, IPC queue pairs, etc.

With the number of VMs increasing due to more powerful processors and larger memories, there is also a potential scaling issue with respect to the number of VMs that can be supported, since each VM requires hardware resources in the NIC.

The management of a VEB in the adapter is also problematic. Typically, the management of servers and the network is split between two separate groups

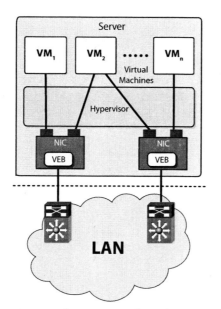

Figure 67: VEB in the adapter

with different capabilities. Placing the VEB in the adapter requires that both groups coordinate to manage the servers, and this is undesirable.

To overcome these limitations the following architecture has been developed.

3.2.5. VEB in the switch

The *"VEB in the switch"* solution delegates complex and performance critical data path functions to an external Ethernet switch, as shown in Figure 68. The Ethernet switches are responsible of ensuring feature consistency to all VMs, independent of where they are located, i.e., of which Hypervisor/Physical-Server the VM resides in.

The number of switches to be managed does not increase, since there are no switches inside the NICs. Also the need for the network managers to manage a server component is no longer present.

NIC designers can now use the available logical gates to provide better performance to ULP, by improving data movement and ULP features like TCP offload, RDMA, and FC/SCSI.

This approach requires developing a new Ethernet tagging scheme between the NIC and the Ethernet switch to indicate the vNIC (virtual NIC) associated within the frame. Cisco Systems has pioneered the VNTag (Virtual NIC

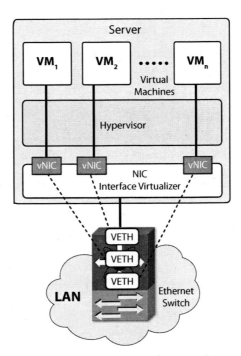

Figure 68: VEB in the switch

Tag) scheme described in Section 3.2.6. This is an IV (Interface Virtualization) scheme that moves the interfaces from the VMs to an external Ethernet switch and makes them virtual. These new virtual interfaces are also called Veth (Virtual Ethernet) ports and the Ethernet switch operates on them if they were physical ports. Whatever an Ethernet switch can do on a physical port, it can do it on a Veth port.

Basically, this new tag binds a vNIC to a Veth port and vice versa. Policies are applied to Veth ports and therefore to vNICs. Policies may include ACLs (Access Control Lists), traffic management (for example, parameters for PFC and ETS), authentication, encryption, etc.

The VNTag is much more difficult to spoof than a MAC Address, since it is inserted by the NIC or the hypervisor and not by the VM.

When a VM moves from one server to another, its vNICs moves with it, the Veth ports moves with their associated policies. This guarantees feature and policy consistency independently of the location of the VM.

This approach requires a control protocol between the NIC and the Ethernet switch which is used to create, assign, modify and terminate the relationship between vNICs and Veths. For example, when a VM moves from Host-A to

Host-B, this control protocol is in charge of terminating the vNICs/Veths relationships on the Ethernet switch where Host-A is connected, and of recreating equivalent relationships on the Ethernet switch where Host-B is connected. Policies associated with the vNICs and Veths are also maintained in the VM migration.

Going back to Figure 68, the NIC contains an "Interface Virtualizer" which is responsible to dispatch frames to/from the vNICs. The Interface Virtualizer is not a switch and therefore it does not allow the direct communication of two vNICs. The interface virtualizer allows:

- one vNIC to communicate with the Ethernet switch;
- the Ethernet switch to communicate with one or more vNICs.

There is no change in the Ethernet switching model; in the presence of Veth port the Ethernet switch has an expanded number of ports, one for each connected vNIC. The Ethernet switch functionality for the virtual ports remains unchanged.

3.2.6. VNTag

The VNTag (Virtual NIC tag) is an Ethernet tag inserted into the Ethernet frame immediately after the MAC-DA (Destination Address) and MAC-SA (Source Address) pair (see Figure 69). The IEEE MACsec (authentication and encryption) tag may precede it.

VNTag is needed to augment the forwarding capability of an Ethernet switch and make it capable to operate in a virtualized environment. Classical Ethernet switches do no support the forwarding of frames where the source and destination MAC address are on the same port and therefore do not support forwarding frames between two VMs connected on the same switch port. VNTag solves this and other issues by creating a virtual Ethernet interface per each VM on the switch. Since the switch is capable of forwarding between these virtual Ethernet interfaces, it is capable of forwarding between VMs connected on the same physical port.

VNTag is a six byte tag whose format is shown in Figure 70. The VNTag is used between the VEB in the Ethernet switch and the Interface Virtualizer. Its main concept is the "vif" (virtual interface identifier). The vif appears in the VNTag as src_vif (source vif) and dst_vif (destination vif).

It starts with two bytes of Ethertype = VNTag in order to identify this particular type of tag. The next four bytes have the following meaning:

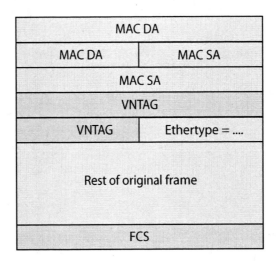

Figure 69: VNTag applied to an Ethernet frame

- v: version[2] – Indicates the version of the VNTag protocol carried by this header, currently version zero.

- r: reserved[1] – This field is reserved for future use.

- d: direction[1] – d = 0 indicates that the frame is sourced from an Interface Virtualizer to the Ethernet switch. d = 1 indicates that the frame is sourced from the Ethernet switch to an Interface Virtualizer (one or more vNICs).

- dst_vif[14] and p: pointer[1] – These fields select the downlink interface(s) that receives a frame when sent from the Ethernet switch to the Interface Virtualizer. As such, they are meaningful when d = 1. When d = 0, these fields must be zero.

- p = 0 indicates that dst_vif selects a single vNIC, typically used for unicast frames. p = 1 indicates that dst_vif is an index into a virtual

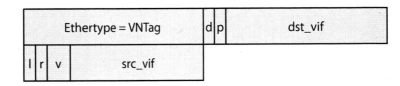

Figure 70: VNTag format

interface list table; this case is used for multicast frame delivery.

- src_vif[12] – Indicates the vNIC that sourced the frame. src_vif must be set appropriately for all frames from an Interface Virtualizer to the Ethernet switch (d = 0). In addition, src_vif may be set for frames headed towards the Interface Virtualizer (d = 1) as indicated by looped.

- l: looped[1] – Indicates that the Ethernet switch looped back the frame towards the Interface Virtualizer that originated it. In this situation, the Interface Virtualizer must ensure that the vNIC indicated by src_vif does not receive the frame, since it has originated the frame.

The vif is just an identifier of the association between a vNIC and a Veth on a given link. It has a meaning that is local to the link (physical port of the Ethernet switch); but it is not unique in the network. If the VM moves, the vNIC and the Veth move with it. The existing association between the vNIC and the Veth on the current link is terminated and a new association is created on the new link. Because vifs are local to a particular link, the vif used to identify the association on the new link may differ from the vif previously used before the move.

The implementation of VNTag can be done either in hardware by a VNTag capable NIC or in software by the hypervisor.

The Interface Virtualizer is commonly found in server NICs, but it can also be implemented as a separate box that acts as a remote multiplexer toward an Ethernet switch. This box is commonly called a "Fabric Extender".

3.2.7. Fabric Extenders

The Fabric Extender is a standalone implementation of an Interface Virtualizer and it uses VNTag toward the upstream Ethernet switches. The Fabric Extender can also be viewed as a remote line-card of an Ethernet switch or as an Ethernet multiplexer/de-multiplexer. These remote line cards are used to increase the number of ports available without increasing the number of management points. Being a simple box that is managed by the upstream switch, it also reduces the overall cost of the solution.

Figure 71 shows three Fabric Extenders connected to two Ethernet switches. Note that the links that cross the dashed line use the VNTag encapsulation.

A Fabric Extender can be a one RU (Rack Unit) box with 48 x 1GE down-

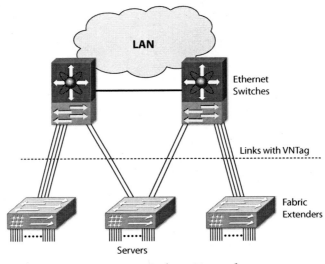

Figure 71: Fabric Extenders

facing ports toward the servers and 4 x 10GE uplinks toward the Ethernet switches. In this particular example the Fabric Extender uses 48 vifs ti identify each 1GE ports and it does not use VNTag on the 1GE interfaces.

The uplinks of the Fabric Extender can be bundled together using EtherChannels. In Figure 71, the left most and right most Fabric Extenders have the four uplinks bundled together toward a single Ethernet switch; the center Fabric Extender has the four uplinks bundled together into a single EtherChannel, but connected in pairs to two separate Ethernet switches.

For example, a data center switch like the Cisco Nexus 5000 can support 12 Fabric Extender providing a total number of 576 x 1GE ports, with a single point of management, in 14 RUs. The Nexus 5000 and attached Fabric Extenders create a single switch domain on the network. The Nexus 5000 performs all the management plane functions and no local configuration or software image is stored on the Fabric Extenders themselves.

Another example is a Fabric Extender with 32 x 10GE downfacing ports toward the servers and 8 x 10GEuplinks toward the Ethernet switches. In this particular example the Fabric Extender uses 256 vifs, and dynamically assign them to the 32 x 10GE ports that also use VNTag on the links toward the servers. This illustrates the possibility to aggregate links that already use VNTag.

Five of these Fabric Extenders, in association with a Nexus 5020, can provide 160 x 10GE ports, with Unified Fabric support, in 7 RUs, with a single point of management.

The Fabric Extender can be placed close to the servers so that low cost copper connections can be used between the servers and the Fabric Extender. Alternatively, the Ethernet switches can be placed at the end-of-row using more expensive fiber connections between the fabric extender and the Ethernet switches. The resulting solution, though oversubscribed, provides a cost-effective 10GE solution that can be centrally managed.

3.2.8. VN-Link

Cisco® Systems and VMware® collaboratively developed the Virtual Network Link (VN-Link) to address the issues of VEBs. VN-Link simplifies the management and administration of a virtualized environment by bringing the server and the network closer together. VN-Link delivers network-wide VM visibility and mobility along with consistent policy management.

During the development of VN-Link a specification has been submitted to the IEEE. It proposes a VEB-in-the-switch architecture that can be implemented either in hardware or in software.

VN-Link and The Nexus 1000v

The Nexus1000V is the first implementation of the VN-link architecture. It is a Cisco software switch embedded into the VMware® ESX hypervisor. It is compliant with VMware® vNDS (virtual Network Distributed Switch) API

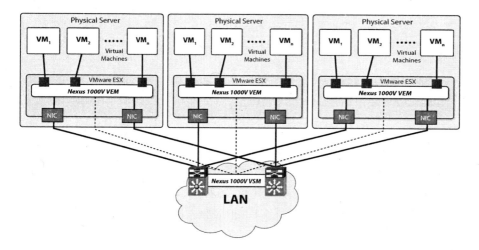

Figure 72: Nexus 1000V

that was jointly developed by Cisco Systems and VMware.

vNetwork DS includes not only the support for VN-Link and VNTag, but also VMDirectPath and NetQueue (see Section 2.5.2.).

VMDirectPath allows VMware® ESX® to bypass the hypervisor and map the physical NICs directly to the virtual machines. This feature is supported, for example, in Cisco Palo, see Section 4.6.

VM policy enforcement is applied to and migrated with the VM when a Vmotion or DRS (Distributed Resource Scheduler) moves a VM. Not only the policies are moved with the VM, but also all the statistical counters, the Netflow status and ERSPAN sessions.

Network Policies are called "Port Profiles" and are created on the Nexus 1000v by the network administrator using, for example, CLI commands. Port Profiles are automatically populated inside VMware® VC (Virtual Center). Port Profiles are visible inside VMware® Virtual Infrastructure Client as "Port Groups" and the server administrator can assign them to vNICs and therefore ultimately to VMs.

Port Profiles are the constructs in VN-Link that enable a collaborative operational model between the server administrator and the network administrator without requiring the use of a new management tool.

Veths and Port Profiles are the basic building blocks to enable automated VM connectivity and mobility of policies, i.e., to allow the interface configuration, interface state and interface statistics to move with a virtual machine from server to server, as VMotion or DRS occurs. This also guarantees that security and connectivity policies are persistent.

Nexus 1000V is a member of the Nexus family of switches and it runs NX-OS. Figure 72 shows the basic components: the "Nexus 1000V VEM" (Virtual Ethernet Module) is installed in VMware® ESX® and remotely configured and managed by the Nexus 1000V VSM (Virtual Supervisor Module) that runs in an NX-OS appliance (virtual or physical) or in a Nexus family switch, like the Nexus 5000 or the Nexus 7000. The dashed lines in Figure 72 indicate the management and configuration relationship between the VEMs and VSM. A single VSM can manage multiple VEMs. In addition it maintains the VMware® Virtual Center provisioning model for server administration.

Being part of the Nexus family, the Nexus 1000V provides many value added features like ACLs, QoS marking and queueing, Cisco TrustSec, CDP v2, Netflow V9, Port Profiles, Cisco CLI, XMP API, SNMP Read/Write, detailed interface counters & statistics, Port Security, etc. It also supports ERSPAN

(Encapsulated Remote SPAN) to allow traffic to mirror the traffic of a VM to an external sniffer located centrally in the data center even during the VM migration.

VN-Link and the Nexus 5000

The Nexus 5000 also supports the VN-Link architecture. It operates as a VEB in the switch (see Section 3.2.5.) using VNTag (see Section 3.2.6.) between the Interface Virtualizer in the hypervisor and the Nexus 5000.

The Interface Virtualizer is in charge of tagging the frames leaving the VM and this can happen in one of two ways:

- inserting the VNTag in the hypervisor software switch (either the Nexus 1000V or the VMware® switch) and then forward the packet to the Nexus 5000, or

- having a NIC (e.g., an SR-IOV NIC) that is capable of adding VN-Tag in hardware. In this second approach, hypervisor bypass can be achieved using an SR-IOV capable NICs.

The external behavior and benefits are the same as using VN-Link on the Nexus 1000V, however with the Nexus 5000 all the Ethernet features (switching, ACLs, ERSPAN, QoS, etc) are performed completely in hardware at wire rate.

Network Policies are still defined in terms of Port Profiles. All VMotion/DRS and VN-Link features are also supported.

3.3.   Memory Expansion

With the growing computing capacity of modern CPUs, memory often becomes the bottleneck limiting server performance. This is true both for memory speed and memory size.

Each server socket has a limited number of memory sockets and speeds with which it can connect to. For example, the Intel® Xeon 5500 (Nehalem-EP) processor used on the first generation of the California system has a supported configuration in which each of the three memory channel per socket is capable of connecting to eight ranks of memory on two DDR3 RDIMMs (see Section 2.2.7.). At the time of writing (end of 2008) the largest RDIMM on the market is 8 GB, therefore 16 GB per bus, and 48 GB per socket.

3.3.1. Speed vs. Capacity

There is a trade-off between speed and capacity. The maximum speed supported varies from 800 Mtps, to 1066 Mtps (Mtps = Million Transfers per second), to 1333 Mtps (see Table 3), depending on how many sockets are installed per memory channel, how many are populated, and how many ranks are present per DIMM.

DIMM slots per channel	DIMM populated per channel	Million Data transfer per second	Ranks per DIMM
2	1	800, 1066, 1333	SR, DR
2	1	800, 1066	QR
2	2	800, 1066,	SR, DR
2	2	800	SR, DR, QR
3	1	800, 1066, 1333	SR, DR
3	1	800, 1066	QR
3	2	800, 1066	SR, DR
3	2	800	SR, DR, QR
3	3	800	SR, DR

SR: Single Rank, DR: Double Rank, QR: Quad Rank

Table 3: Trade-off of DIMM speed vs. capacity

3.3.2. Capacity vs. Cost

Another important consideration is the cost of the DIMM per GB. This cost does not grow linearly between generations. For example, the prices shown in Table 4 are the list prices at the time this book was written.

Capacity	List Price	List Price / GB
2 GB	125 USD	65.5 USD/GB
4 GB	300 USD	75 USD/GB
8 GB	1,200 USD	150 USD/GB

Table 4: DDR3 DIMM list price

While there is not much difference in the price/GB using 2 or 4 GB RDIMMs, 8 GB RDIMMs come at a substantial premium.

3.3.3. How much memory is required?

Memory requirements are hard to quantify in absolute terms, but guidelines can be given for applications.

Memory on Desktop/Laptop

To start consider the typical memory requirements shown in Table 5 for operating systems for desktops and notebooks.

Operating System	Typical	High Performance
Windows Vista	1GB – 4 GB	2GB – 8GB
Windows XP	512MB – 2GB	1GB – 4GB
Windows 2000	256MB – 1GB	512MB – 2 GB
Linux	512MB – 2GB	1GB – 4GB
Macintosh OS X	512MB – 2GB	1GB – 4GB

Table 5: OS memory requirements

These numbers may not look too bad at first compared to the capacity of current DIMM. When considered in a Virtual Machine environment, they get multiplied by the number of VMs and they often exceed the capacity of current DIMMs, before the CPU becomes a bottleneck (more on this later).

Memory on servers

When considering server memory requirements, the goal is to have enough physical memory compared to active virtual memory, so that memory pagination can be avoided. Virtual memory is paged to disk and this is a slow process. If intense pagination occurs, significant performance degradation is unavoidable.

Memory also depends on the number and speed of processors. A server with more cores processes information faster and requires more memory than a server with fewer cores.

64-bit operating systems also use more memory than their 32-bit predecessors.

Memory requirements, while clearly dependent on the number of users, depend more heavily on the type of applications being run. For example, text based applications tend to require less memory than GUI/Graphic-based applications.

The applications that tend to use less memory are applications like directory services (DNS, DHCP or LDAP servers) that perform a very specific task. 1GB to 2GB can be reasonable.

Communication servers (like IP phone, voice servers, Email/Fax servers) are intermediate memory users in the 2GB to 4GB range.

Web server memory mainly depends on the number of concurrent users and the required response time: they are in the 2GB to 4GB range.

Application server memory is in a similar range as web servers and it greatly depends on the application and on the number of users.

Database servers are in general the most memory intensive and can use the maximum amount of memory a server can provide to cache the database content thus dramatically improving performance. They start at 4GB for small databases.

Some engineering tools are memory-bound on the complexity of the model they can simulate; the larger the memory on the server the larger the circuit they can simulate or the more accurate the analysis they perform. In the same class are tools for oil and energy and biotech research.

Memory on Virtualized Servers

The amount of memory installed in a server may be insufficient in a virtualized environment where multiple virtual machines run on the same processor. Each virtual machine consumes memory based on its configured size, plus a small amount of additional overhead memory for virtualization [20].

Hypervisors typically allow specifying a minimum and a maximum amount of memory per VM: the minimum is guaranteed to be always available; the maximum can be reached as a function of the state of the other VMs. This technique is called memory over-commitment and it is useful when some virtual machines are lightly loaded while others are more heavily loaded, and relative activity levels vary over time.

Certain minimums must be guaranteed to avoid errors. For example Windows® XP and RedHat® Enterprise Linux 5 both requires at least 512MB.

In practice most production environments allocate from 2GB/VM to 4GB/ per VM. Using this assumption, Table 6 analyzes how many VMs can be run on a Nehalem CPU socket assuming 2 DIMMs per memory channel are populated.

	Total Memory	2GB/VM	4GB/VM	Estimated list price	List price per 2GB/VM
Inexpensive 2 GB DIMM	12 GB	6	3	750 USD	125
Medium price 4 GB DIMM	24 GB	12	6	1,800 USD	150
Expensive 8GB DIMM	48 GB	24	12	7,200 USD	300

Table 6: VMs per socket

This table illustrates how either expensive DIMMs are needed to run a significant number of VMs or few VMs can be run per socket with the possibility of not fully utilizing all the cores. For example, 3 VMs will not fully utilize the 4 cores of a Nehalem socket and even 6 or 12 VMs may not be enough, unless they run a CPU intensive workloads.

In the Intel® Microarchitecture (Nehalem), memory controllers are integrated on the processor socket and therefore memory can only be added with additional sockets, but this does not solve the original issue of increasing the memory per socket.

3.3.4. NUMA

Non-Uniform Memory Access or Non-Uniform Memory Architecture (NUMA) is a computer memory design used in multiprocessors, where the memory access time depends on the memory location relative to a processor.

Under NUMA, a processor can access its own local memory faster than non-local memory, that is, memory local to another processor or memory shared between processors [31].

The Intel® Xeon 5500 (Nehalem-EP) processor family supports NUMA and allows a core to access the memory locally connected to the three memory channel of its socket or to access the memory connected to another socket via the QPI interconnect. Access times are of course different, for example the memory local to the processor socket can be accessed in ~63 nanoseconds, while the memory connected to another processor socket can be accessed in ~102 ns.

The particular type of NUMA supported by Nehalem is "cache-coherent NUMA", i.e., the memory interconnection system guarantees that the memory and all the potentially cached copies are always coherent.

The Nehalem support of NUMA is a key enabling technology for the California solution described in the next section.

3.3.5. *The California approach*

Project California addresses the need for larger memories with dedicated computing blades that have an increased number of memory sockets. This may seem in contradiction with what shown in Table 3, where the limitation in number of memory modules is reported.

The limit on the number of memory sockets is not directly related to the addressing capability of the processor (Nehalem is capable of addressing larger memories). Electrical issues, DRAM device density, and details of the processor's pinout are the real limiting factors. Each memory socket on a memory channel adds parasitic capacitance, and the DIMMs inserted on the memory sockets further load the memory bus. Increased the load results in lower speed operation. Three DIMMs directly attached on the bus is the practical upper limit given the speeds required for DDR3 operation. Additionally, memory capacity is limited by the maximum number of ranks (eight) that the Nehalem processor can directly address on each memory channel, and by the density of the DRAM devices that comprise each rank. Finally, there are details in the number and type of control signals generated by the Nehalem CPU that limit the number of directly attached DIMMs to two or three.

When a larger memory system is desired, California uses Cisco ASICs called the "Catalina chipset" to expand the number of memory sockets that can be connected to each single memory bus. These ASICs are inserted between the processor and the DIMMs on the memory bus, minimizing the electrical load, and bypassing the control signal limitations of the Nehalem CPU design.

The achieved expansion factor is 4X: using 8GB DIMMs a Nehalem socket can connect up to 192 GB of RAM and a dual processor socket blade can host up to 384 GB of RAM (see a Windows screenshot in Figure 73). The DIMMs are unmodified JEDEC standard DDR3 RDIMMs. This expansion is done at the electrical level and it is completely transparent to the operating systems and its applications. The BIOS is extended to initialize and monitor the ASICs and to perform error reporting.

In order to increase the number of memory sockets without sacrificing memory bus clock speed, the Catalina chipset adds a small amount of latency to the first word of data fetched from memory. Subsequent data words arrive at the full memory bus speed with no additional delay. For example in a typical Nehalem configuration the memory latency grows from ~63 nanoseconds to ~69 nanoseconds (+10%), but it is still well below the 102 nanoseconds (+62%) required to access the memory on a different socket, and is orders of magnitude below the time required to fetch data from disk.

Figure 74 illustrates how each memory channel is buffered and expanded to 4 sub-channels.

Each sub-channel can support 2 SR (Single Rank), DR (Dual Rank) or QR (Quad-Rank) x4 DIMMs. "x4" refers to the data width of the DRAM chips

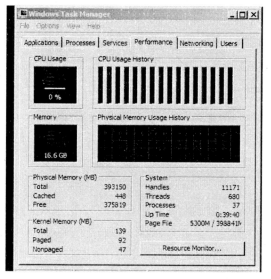

Figure 73: Windows running with 384 GB of RAM

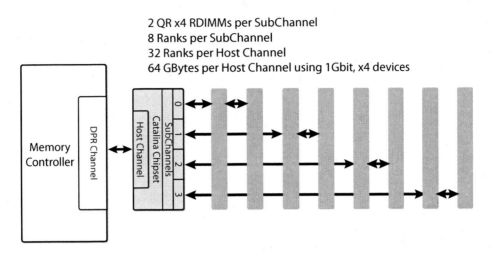

Figure 74: The Catalina ASIC chipset

in bits.

Figure 75 and Figure 76 show a comparison between a classical two-socket Nehalem system and a two-socket expanded system.

3.3.6. The California advantage

California with the adoption of the Catalina chipset, multiplies the number

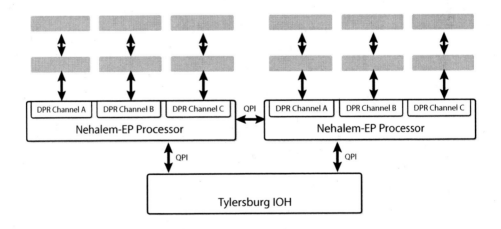

Figure 75: Nehalem system without memory expansion

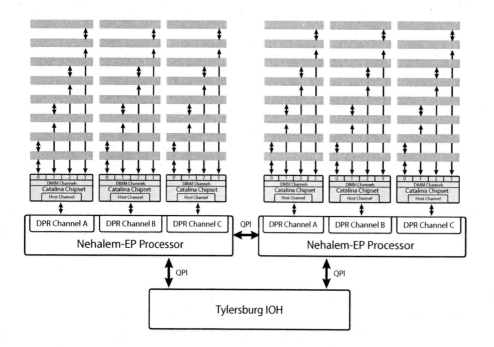

Figure 76: Nehalem system with memory expansion

of memory sockets available by four while introducing minimal latency. This solution uses standard DDR3 DIMMs.

This can be exploited in two different ways:

- building the largest memories possible using the highest density DIMMs. This may be particularly appropriate for applications that are more memory intensive, such as large databases and engineering tools like large circuit synthesis and simulation.

- building a medium to large memory configuration using inexpensive DIMMs.

In both cases the final customer obtains the best price/performance ratio.

4.   I/O Adapters

Historically, server I/O connectivity to storage, local area networks, and other servers have been accommodated by hardware and software specifically designed and dedicated to the application, each with different traffic characteristics and handling requirements. Server access to storage in enterprises has been largely satisfied by Storage Area Networks (SAN) based on Fibre Channel. Server access to networking devices is almost exclusively satisfied by local networks (LAN) based on Ethernet. Lastly, server-to-server connectivity, or inter-process communication (IPC) for high performance computing (HPC) is being satisfied with InfiniBand. While each of these technologies serves their application needs very well, the specialization results in the need for enterprise servers to support multiple dedicated specialized networks with unique dedicated hardware. All these networks in the data center often require physically redundant backup connections which compounds the problem and introduces double the amount of NICs, HBAs & HCAs into a the network.

In recent years, server density has increased and server manufacturers are providing more compute power in smaller and smaller packaging. "Green" concerns (see Section 1.1.1.) have sparked a strong emphasis on reduced power consumption and cooling. Servers one "RU" in height and bladed chassis-based servers are pushing the density envelope. The trend toward higher and higher densities is resulting in tremendous incentive to provide consolidated I/O. Consolidated I/O can result in fewer unique I/O adapters, translating to less power consumption and heat dissipation, as well as reduced, simplified, common cabling[1].

Players in this new consolidated I/O market are the classical Ethernet NIC vendors that are adding storage capabilities to their devices and HBA vendors that are adding Ethernet support.

Project California uses several different I/O adapters, all in mezzanine form factor. This chapter presents some of them that will be available at first customer shipment or future releases. Please consult the California documentation to verify availability.

4.1.   Disclaimer

The material contained in this chapter has been provided by the I/O adapter

---

[1]   The authors thank QLogic® for providing most of the text of this section

manufacturers. For this reason the presentation of the different I/O adapters are not homogeneous and even the terminology used differ from adapter to adapter. The authors understand that this can cause some confusion in the reader.

4.2.   The Intel® approach

Virtualization focuses on increasing service efficiency through flexible resource management. In the near future, this usage model will become absolutely critical to data centers, allowing IT managers to use virtualization to deliver high availability solutions with the agility to address disaster recovery and real-time workload balancing so they can respond to the expected and the unexpected[2].

Dynamic load balancing requires the ability to easily move workloads across multiple generations of platforms without disrupting services. Centralized storage becomes a key requirement for virtualization usage models. 10GE (10 Gigabit Ethernet) supports storage implementations of iSCSI SAN as well as NAS today. Fibre channel over Ethernet (FCoE) solutions extend the 10GE to seamlessly interoperate with existing Fibre Channel SAN installations. Consolidating storage traffic over 10GE using FCoE simplifies the network and provides a cost effective mean to enable virtualization usage.

In addition to virtualization there are many bandwidth hungry applications that require 10GE NICs. 10 Gigabit Ethernet provides the bandwidth needed by High Performance Computing (HPC) applications, clustered databases, image rendering, and video applications.

4.2.1.   10 Gigabit Ethernet NIC solutions

Intel® launched one of the first PCI-X based 10GE Adapters to the market back in 2003. Since then new PCIe based products have been introduced (see Figure 77).

Today, the 10GE NIC market is predominately served with stand-alone adapter cards. These cards plug into the PCIe (PCI Express, see Section 2.3.1.) bus expansion slots of a Server and provide 10GE connectivity.

Mezzanine form-factor NICs are appropriate for server blades that are installed into blade servers. In this case there is not a standard PCIe slot. The mezzanine

---

2   The authors thank Intel® for providing the text of the next few sections. The pictures are courtesy of Intel®.

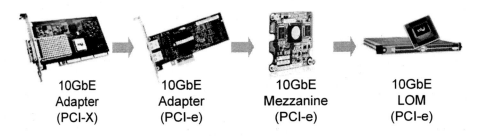

| 10GbE
Adapter
(PCI-X) | 10GbE
Adapter
(PCI-e) | 10GbE
Mezzanine
(PCI-e) | 10GbE
LOM
(PCI-e) |

Figure 77: NIC evolution

is a daughter card with a proprietary connector, size, and power consumption. While the connector is proprietary, the protocol spoken is always PCIe.

In the next couple of years, OEMs will start providing 10GE LAN on Motherboard (LOM) solutions. LOM solutions implement LAN chipset integrated onto the PC motherboard thus freeing up valuable PCI slots. LOM implementations are less expensive than adapter or mezzanine cards and also implement effective management technologies that require signal transfer directly to the system board. With LOM implementation 10GE will be pervasive; every server will ship with an on-board 10GE controller.

The tipping point for 10GE will come when data centers refresh their environment with the new Intel® Xeon® 5500 class of server products (formerly known by the codename Nehalem, see Section 2.4.).

The Figure 78 shows that previous generation platforms were saturated at 2 ports 10GE throughput. The Intel® Nehalem platform is able to scale up to five 10GE ports and beyond and still have enough CPU headroom for other tasks.

Two specific Intel® 10 Gigabit Adapters that can be used in the California platform are presented in the next sections, outlining advantages and peculiarities.

4.3. Intel® 82598 10 Gigabit Ethernet Controller (Oplin)

Intel® 82598 is a high-performing, PCIe 1/10 Gigabit Ethernet controller that is ideally suited for demanding enterprise applications and embedded system designs. It includes support for multi-core processors, it improves virtualized server performance, and it delivers efficient storage over Ethernet. The 82598

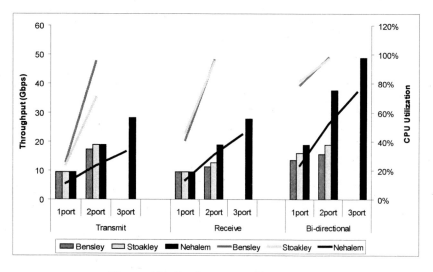

Figure 78: Performance Comparison

Source: Intel® Labs 2008. Performance calculated based on 64k I/O IxChariot test with Intel® 82598 10 Gigabit Ethernet controller, with Linux 2.6.18 kernel on a Quad Core Intel® Xeon® Platform (formerly Bensley/Stoakley) and Intel® Next Generation (Nehalem). Actual performance may vary depending on test and system configuration.

is an ideal candidate for server consolidation. It is a single chip dual port 10GE implementation in a 31x31mm package (see Figure 79) with industry leading power consumption of 4.8 Watts (typical).

Project California supports a mezzanine card named Cisco UCS 82598KR-CI that is based on the Intel® 82598,

Figure 80 shows a block diagram of the 82598 that integrates two 10 Gigabit Ethernet Media Access Control (MAC) and XAUI ports. Each port supports CX4 (802.3ak) and KX4/KX (802.3ap) interfaces and contains a Serializer-Deserializer (SerDes) for backward compatibility with gigabit backplanes. The device is designed for high performance and low memory latency. Wide internal data paths eliminate performance bottlenecks by efficiently handling large address and data words. The controller includes advanced interrupt-handling features and uses efficient ring-buffer descriptor data structures, with up to 64 packet descriptors cached on a chip. A large on-chip packet buffer maintains superior performance. The communication to the Manageability Controller (MC) is available either through an on-board System Management BUS (SMBus) port or through the Distributed Management Task Force (DMTF)

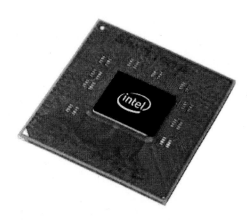

Figure 79: The Intel® 82598 Oplin ASIC

defined NC-SI. The 82598 also supports PXE, and iSCSI boot.

The 82598 was designed to offer outstanding performance and power efficiency, while at the same providing support for multi-core processors, virtualization and unified I/O over a broad range of Operating Systems and virtualization hypervisors. It can be installed in PCIe form-factor cards, or in mezzanine cards or directly on the server motherboard (LOM).

Figure 81 shows the 82598 in mezzanine form factor for California.

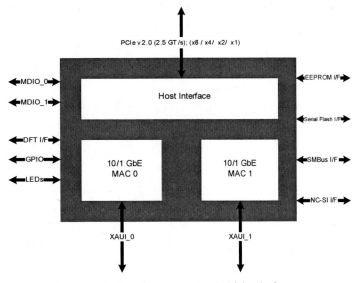

Figure 80: Intel® 82598 Oplin block diagram

4.3.1. *Support for multi-core CPUs*

The 82598 introduces new networking features to support multi-core processors. Supported features include:

- Message Signaled Interrupts-Extended (MSI-X) that distributes network controller interrupts to multiple CPUs and cores. By spreading out interrupts, the system responds to networking interrupts more efficiently, resulting in better CPU utilization and application performance.

- Multiple Tx/Rx queues, a hardware feature that segments network traffic into multiple streams that are then assigned to different CPUs and cores in the system. This enables the system to process the traffic in parallel for improved overall system throughput and utilization.

- Low latency that enables the server adapter to run a variety of protocols while meeting the needs of the vast majority of applications in HPC clusters and grid computing.

- Intel® QuickData Technology, which enables data copy by the chipset instead of the CPU, and Direct Cache Access (DCA) that enables the CPU to pre-fetch data, thereby avoiding cache misses and improving application response times.

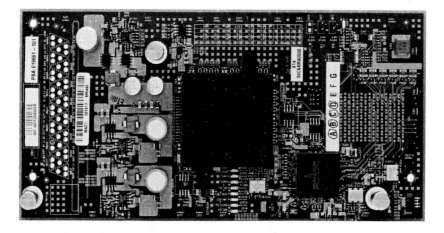

Figure 81: Cisco UCS 82598KR-CI using Intel® 82598

4.3.2. *Hardware-assisted Virtualization*

Intel® has invested in hardware support for improving virtualization performance. These hardware enhancements have been implemented in the processors, in the chipsets and in the adapters. The 82598 includes support for the Intel® Virtualization Technology for Connectivity (Intel® VT-c), to reduce the need for compute-intensive software translations between the guest and host OSes. An important assist for virtualized environments is the Virtual Machine Data Queues (VMDq), which is described in Section 2.5.1.

4.3.3. *Advanced Features for Storage over Ethernet*

The 82598 includes support for iSCSI acceleration and advanced features for unified storage connectivity. It supports native iSCSI storage with all major operanting systems including Microsoft®, Linux®, and VMware®, as well as iSCSI remote boot.

An advantage of iSCSI compared to FC is cost: since iSCSI can run over existing IP infrastructure (as oppose to a dedicated FC network) this can provide the storage features at a reduced price.

iSCSI remote boot offers a number of advantages and simplifications: Server Consolidation and Virtualization. Remote boot enables servers to boot from an OS image on the SAN. This is particularly advantageous for servers in high-density clusters as well as for server consolidation and virtualization.

With OS images stored on the SAN, provisioning new servers and applying upgrades and patches is easier to manage; when a server remotely boots from the SAN, it automatically acquires the latest upgrades and fixes.

Improved Disaster Recovery-All information stored on local SAN-including boot information, OS image, applications, and data can be duplicated on a remotely located SAN for quick and complete disaster recovery.

4.4.   Intel® 82599 10 Gigabit Ethernet Controller (Ninatic)

The 82599 is the second-generation PCIe Gen2 based 10GE controller from Intel. The 82599 delivers enhanced performance by including advanced scalability features such as Receive Side Coalescing and Intel® Flow Director Technology. The 82599 delivers virtualization capabilities implementing Single Route I/O virtualization (SRIOV, see Section 3.2.2.). Finally, the 82599 delivers advanced capabilities for Unified network including support for iSCSI, NAS, and FCoE.

At the time of writing this controller is not yet shipping, so there is not yet an announced mezzanine for California.

Like the 82598, the 82599 is a single chip, dual port 10GE implementation in a 25x25mm package (see Figure 82). It also integrates serial 10GE PHYs and provides SFI and KR interfaces. With a power consumption of less than 6 Watts (typical[3]), a small footprint, and integrated PHYs, the 82599 is suited for LOM and Mezzanine card implementations. The advanced features of 82599 along with the Intel® Nehalem processor enables customers to scale volumes servers to fully utilize and scale to 10GE capacity.

Figure 83 shows a 82599 that integrates two 10 GbE Media Access Control (MAC) and Multi-Speed Attachment Unit Interface (MAUI) ports that support IEEE 802.3ae (10Gbps) implementations as well as perform all of the functions called out in the standards for:

- XAUI;

- IEEE 802.3ak and IEEE 802.3ap backplane Ethernet (KX, KX4, or KR);

- PICMG3.1 (BX only) implementations including an auto-negotiation layer and PCS layer synchronization;

- SFP+ MSA (SFI).

Figure 82: Intel® 82599 10 Gigabit Ethernet Controller

---

3   Note that these power consumption ranges are interface-dependent.

4.4.1. Improved performance

The 82599 is designed with improved data throughput, lower memory latency and new security features. Performance improvements include:

- Throughput — the 82599 provides wire speed dual-port 10 Gbps throughput. This is accomplished using the PCIe physical layer (PCIe Gen 2), by tuning the internal pipeline to 10 Gbps operation, and by enhancing the PCIe concurrency capabilities.

- Latency — reduced end-to-end latency for high priority traffic in presence of other traffic. Specifically, the 82599 reduces the delay caused by preceding TCP Segmentation Offload (TSO) packets. Unlike previous 10 GbE products, a TSO packet might be interleaved with other packets going to the wire.

- Tx Descriptor Write Back — the 82599 improves the way Tx descriptors are written back to memory. Instead of writing back the Descriptor Done (DD) bit into the descriptor location, the head pointer is updated in system memory.

- Receive Side Coalescing (RSC) — RSC coalesces incoming TCP/ IP (and potentially UDP/IP) packets into larger receive segments. It is the inverse operation to TSO on the transmit side. The 82599

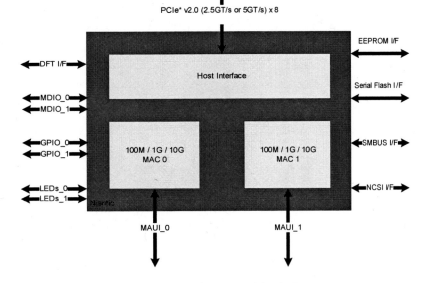

Figure 83: Intel® 82599 block diagram

can handle up to 32 flows per port at any given time.

- PCIe v2.0 — Several changes are defined in the size of PCIe transactions to improve the performance in virtualization environment.

- Rx/Tx Queues and Rx Filtering — the 82599 Tx and Rx queues have increased in size to 128 Tx queues and 128 Rx queues. Additional filtering capabilities are provided based on L2 Ethertype, 5-tuples, SYN identification.

- Flow Director — a large number of flow affinity filters that direct receive packets by their flows to queues for classification, load balancing, and matching between flows and CPU cores.

- Interrupts — a new interrupt scheme is available in the 82599:

  - Control over the rate of Low Latency Interrupts (LLI);

  - Extensions to the filters that invoke LLIs;

  - Additional MSI-X vectors.

- Packet Filtering and Replication — the 82599 adds additional coverage for packet filtering for virtualization by supporting the following filtering modes:

  - Filtering by unicast Ethernet MAC address;

  - Filtering by VLAN tag;

  - Filtering of multicast Ethernet MAC address;

  - Filtering of broadcast packets.

- For each of the above categories, the 82599 can replicate packets to multiple Virtual Machines (VMs). Various mirroring modes are supported, including mirroring a VM, a Virtual LAN (VLAN), or all traffic into a specific VM.

- Packet Switching — the 82599 forwards transmit packets from a transmit queue to an Rx software queue to support VM-VM communication. Transmit packets are filtered to an Rx queue based on the same criteria as packets received from the wire.

- Traffic Shaping — transmit bandwidth is allocated among the virtual interfaces to avoid unfair use of bandwidth by a single VM. Allocation is done separately per traffic class (see Section 3.1.6.) so that bandwidth assignment to each traffic class is partitioned among the VMs.

4.4.2. Hardware-assisted Virtualization

The 82599 supports two modes of operation for virtualized environments:

1. Direct assignment of part of the port resources to different guest OSes using the PCI SIG SR-IOV standard. Also known as Native mode or Pass Through mode.

2. Central management of the networking resources by an I/O Virtual Machine (IOVM) or by the Virtual Machine Monitor (VMM). Also known as software switch acceleration mode. This mode is referred to as VMDq mode.

The virtualization offload capabilities provided by 82599, apart from the replication of functions defined in the PCI SIG IOV specification, are part of VMDq.

A hybrid model, where some of the VMs are assigned a dedicated share of the port and the rest are serviced by an IOVM is also supported. However, in this case the offloads provided to the software switch might be more limited. This model can be used when parts of the VMs run operating systems for which VF (Virtual Function, aka vNIC) drivers are available and thus can benefit from an IOV and others that run older operating systems for which VF drivers are not available and are serviced by an IOVM. In this case, the IOVM is assigned one VF and receives all the packets with Ethernet MAC addresses of the VMs behind it.

The 82599 also supports the PCI-SIG single-root I/O Virtualization Initiative (SR-IOV) including the following functionality:

-   Replication of PCI configuration space;
-   Allocation of BAR (Base Address Register) space per virtual function;
-   Allocation of requester ID per virtual function;
-   Virtualization of interrupts.

The 82599 provides the infrastructure for direct assignment architectures through a mailbox mechanism. Virtual Functions (VFs) might communicate with the Physical Function (PF) through the mailbox and the PF can allocate shared resources through the mailbox channel.

4.4.3. Support for DCB (Data Center Bridging)

DCB is a set of features that improve the capability of Ethernet to handle multiple traffic types (see Section 3.1.).

The layer 2 features of DCB implemented in the 82599 are:

- Multi-class priority arbitration and scheduling — The 82599 implements an arbitration mechanism on its transmit data path. The arbitration mechanism allocates bandwidth between traffic classes (TC) in bandwidth groups (BWGs) and between Virtual Machines (VMs) or Virtual Functions (VFs) in a virtualization environment.

- Class-based flow control (PFC — Priority Flow Control) — Class-based flow control functionality is similar to the IEEE802.3X link flow control. It is applied separately to the different TCs.

- DMA queuing per traffic type — Implementation of the DCB transmit, minimization of software processing and delays require implementation of separate DMA queues for the different traffic types. The 82599 implements 128 descriptor queues in transmit and 128 descriptor queues in receive.

- Multiple Buffers — The 82599 implements separate transmit and receive packet buffers per TC.

- Rate-limiter per Tx queue — limiting the transmit data rate for each Tx queue. Rate limiting is part of the BCN (Backward Congestion Notification) mechanism to resolve congestion in the network.

4.4.4. Storage over Ethernet

Refer to the 82598 description for details in Section 4.3.3.

4.4.5. Fibre Channel over Ethernet (FCoE)

Existing FC HBAs offload the SCSI protocol to maximize storage performance. In order to compete with this market, the 82599 offloads the main data path of SCSI Read and Write commands.

82599 offloads FC CRC check, receive coalescing and Direct Data placement (DDP) tasks from the CPU while processing FCoE receive traffic.

FC CRC calculation is one of the most CPU intensive tasks. The 82599 offloads the receive FC CRC integrity check while tracking the CRC bytes and FC padding bytes. The 82599 recognizes FCoE frames in the receive data path by their FCoE Ethernet type and the FCoE version in the FCoE header.

The 82599 can save a data copy by posting the received FC payload directly to the kernel storage cache or the user application space.

When the packet payloads are posted directly to user buffers their headers might be posted to the legacy receive queues. The 82599 saves CPU cycles by reducing the data copy and also minimizes CPU processing by posting only the packet headers that are required for software.

4.4.6. Time Sync — IEEE 1588

The IEEE 1588 International Standard lets networked Ethernet equipment synchronize internal clocks according to a network master clock. The PTP (Precision Time Protocol) is implemented mostly in software, with the 82599 providing accurate time measurements of special Tx and Rx packets close to the Ethernet link. These packets measure the latency between the master clock and an end-point clock in both link directions. The endpoint can then acquire an accurate estimate of the master time by compensating for link latency.

The 82599 provides the following support for the IEEE 1588 protocol:

- Detecting specific PTP Rx packets and capturing the time of arrival of such packets in dedicated CSRs (Control and Status Registers);
- Detecting specific PTP Tx packets and capturing the time of transmission of such packets in dedicated CSRs;
- A software-visible reference clock for the above time captures.

4.4.7. Double VLAN

The 82599 supports a mode where all received and sent packets have an extra VLAN tag in addition to the regular one. This mode is used for systems where the switches add an additional tag containing switching information.

When a port is configured to double VLAN, the 82599 assumes that all packets received or sent to this port have at least one VLAN. The only exception to this rule is flow control packets, which don't have a VLAN tag.

4.4.8. Security

The 82599 supports the IEEE P802.1AE LinkSec specification. It incorporates an inline packet crypto unit to support both privacy and integrity checks on a packet per packet basis. The transmit data path includes both encryption and signing engines. On the receive data path, the 82599 includes both decryption and integrity checkers. The crypto engines use the AES GCM algorithm, which is designed to support the 802.1AE protocol. Note that both host traffic and manageability controller management traffic might be subject to authentication and/or encryption.

The 82599 supports IPsec offload for a given number of flows. It is the operating system's responsibility to submit (to hardware) the most loaded flows in order to take maximum benefits of the IPsec offload in terms of CPU utilization savings. Main features are:

- Offload IPsec for up to 1024 Security Associations (SA) for each of Tx and Rx;

- AH and ESP protocols for authentication and encryption;

- AES-128-GMAC and AES-128-GCM crypto engines;

- Transport mode encapsulation;

- IPv4 and IPv6 versions (no options or extension headers).

4.5. Converged Network Adapters (CNAs)

Converged Network Adapters is a term used by some vendors to indicate a new-generation of consolidated I/O adapters that include features previously present in HBAs, NICs, and HCAs (see also Section 3.1.7.)[4].

CNAs offer several key benefits over conventional application unique adapters, including

- Fewer total adapters needed;

- Less power consumption and cooling;

- Reduced and simplified cabling.

It is important to note that CNAs rely on two distinct device drivers at the OS level: one for the FC part, one for the Ethernet part. This is exactly similar to the way physically separate FC and Ethernet adapters function. This key attri-

---

4   The authors are thankful to QLogic® Corporation for writing most of the text of this section.

bute allows the implementation of CNAs to be evolutionary in nature, while providing revolutionary advances.

CNAs are typically dual-port 10 GE, providing more than adequate bandwidth and high-availability to operate in a converged environment.

The first CNAs to appear on the market were PCIe standard form-factor adapters for industry standard servers, but the same consolidation benefits extend to blade servers using custom form-factor mezzanine cards provided by the server manufacturer.

PCIe form-factor CNAs are appropriate for industry standard servers (e.g., rack mounted servers) that have PCIe slots that use standard PCIe connectors (see Figure 33). The PCIe standards define the size, connector, and power consumption of these adapters guaranteeing multi-vendor interoperability.

Mezzanine form-factor CNAs are appropriate for server blades that are installed into blade servers. A blade server that requires 10 or more dual-port Ethernet (LAN) mezzanine cards and 10 or more dual-port Fibre Channel (SAN) mezzanine cards (one each per processor blade) could reduce the hardware 50% by using 10 or more CNA mezzanine cards instead (one per processor blade). The use of CNAs in the blade server can also reduce the number of embedded switches or pass-through modules required by 50%. This will also significantly simplify backplane complexity and power/thermal burden for blade servers.

The minimum characteristics required for a CNA to operate optimally in a converged environment are listed below:

- Dual-port, full-bandwidth 10 Gbps per port maximum throughput for high bandwidth storage (SAN) and networking (LAN) traffic;

- Hardware offload for FCoE protocol processing;

- Support for de facto industry standard Fibre Channel software stacks;

- Full support for TCP/IP and Ethernet performance enhancements such as priority-based flow control (802.1 Qbb), Enhanced Transmission Selection (802.1Qaz), DCBX protocol (802.1Qaz), jumbo frames, checksum offloads, and segmentation offloads;

- Boot from LAN / SAN: PXE boot, BIOS, UEFI, FCode, depending on OS;

- PCIe 1.0 x8 or PCIe 2.0 x4 system bus support.

Several different companies manufacture CNAs, including Broadcom®, Brocade®, Cisco®, Emulex®, Intel®, and QLogic®.

The next sections describes different mezzanine CNAs that can be used in the California platform, outlining advantages and peculiarities.

4.6.   Cisco® Palo

Palo is a CNA developed by Nuova Systems, now part of Cisco® Systems. It is currently used only on the California platform and therefore it exists only in the mezzanine form factor used by California computing blades. It derives its name from the City of Palo Alto, California. Figure 84 shows a picture of the mezzanine board with Palo called Cisco UCS VIC M81KR Virtual Interface Card.

It is a single chip ASIC designed in 90 nm silicon technology with seven metal layers. The die is approximately 12x12 mm and it is mounted in a 31x31mm FC BGA (Ball Grid Array) package, with 900 balls.

The single ASIC design has major advantages in term of space, cost and power consumption. The Palo ASIC typically has a power consumption of 8 Watts and the overall mezzanine card of 18 Watts, well below the maximum of 25 Watts established by the PCIe standard.

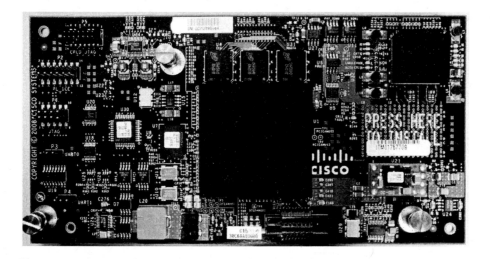

Figure 84: Palo mezzanine board for California

Palo is compliant with PCIe version 1.1, it has 16 lanes for a total of 32 Gb/s of raw bandwidth, full duplex, more than enough to support the two 10 GE ports that provide the external connectivity.

Palo was designed from day one to be a CNA capable of providing hardware support for virtualization. It is capable of handling all traffic to and from the host, including network, storage, inter-processor communication, and management. Management agents run on embedded processors inside Palo, providing server, network, and storage management functions.

When used in conjunction with the California blade, Palo provides several important system-level features:

- I/O consolidation and Unified Fabric;

- DCB compliant Ethernet interfaces;

- Capability to support active/active and active/standby;

- Capability to be fully configured through the network;

- Capability to create I/O devices on demand to better support virtualization;

- Capability to support kernel and hypervisor bypass;

- Low latency and high bandwidth;

- SRIOV compliant hardware;

- Native support for VNTag;

- Native support for FCoE.

The logical diagram of Palo is reported in Figure 85, where the two DCE interfaces are 10 GE interfaces.

Palo has an internal switch that allows it to work as a VEB (Virtual Ethernet Bridge) in the adapter (see Section 3.2.4.) or as a VEB in the switch through the use of VNTag (see Section 3.2.5. and Section 3.2.6.).

The MCPU (management CPU) is in charge of the configuration of Palo and creates the vNIC that are seen by the OS or the hypervisor as regular PCI devices. Palo supports up to 128 vNICs and each vNIC may be either an Ethernet NIC or a FC HBA. Other types of vNICs are supported in hardware, but currently not used.

In the preferred way of operation Palo assigns a VNTag to each vNIC and uses an external switch to switch between them. VNTag also provides traffic separation between the different VMs that share the same Palo adapter.

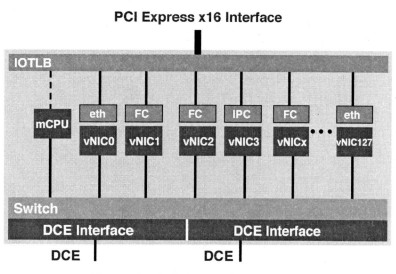

Figure 85: Palo Logical Diagram

This architecture is aligned with VMware technologies known as PTS (Pass Through Switching) and UPT (Uniform Pass Through). In these architectures each VM, for each Ethernet vNIC, gets a data plane interface into the adapter and performs I/O operation without going through the hypervisor. Of course the pass through interface is protected to prevent each VM from interfering with other VMs and with the hypervisor.

Figure 86 details this architecture and shows the strong capability of PCIe virtualization. In PCI terminology a "PCIe device" can be either a traditional endpoint, such as a single NIC or HBA, or a switch used to build a PCIe topology. A PCIe device is typically associated with a host software driver; therefore each Palo entity that requires a separate host driver is defined as a separate PCI device. Every PCIe device has an associated configuration space that allows the host software to:

- Detect PCIe devices after reset or hot plug events;
- Identify the vendor and function of each PCIe device;
- Discover and assign system resources to each PCIe device needs, such as memory address space and interrupts;
- Enable or disable the PCIe device to respond to memory or I/O accesses;
- Tell the PCIe device how to respond to error conditions;
- Program the routing of PCIe device interrupts.

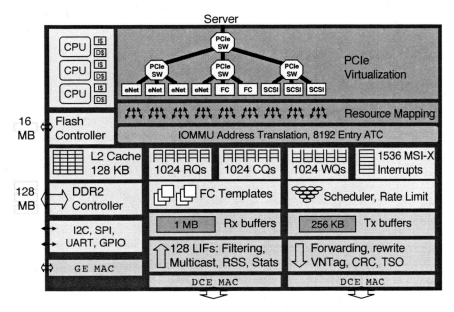

Figure 86: Palo detailed block diagram

Each PCIe device is either type 0 or type 1. Type 0 devices are endpoints, including vNICs. Type 1 devices include switches and bridges, and may also contain additional functions beyond the basic bridge or switch function. For example, the IOMMU control interface is not an endpoint, but rather a function attached to the primary bridge in Palo.

The firmware and configuration of Palo are stored in an external Flash memory. At boot time, the management CPU loads the firmware and the configuration and builds a PCIe topology. This initial topology can be modified later by using the PCIe hot plug interface standard, wherein devices can be removed and added from a running system. The BIOS and the OS/hypervisor see any configured PCIe device on the PCIe bus. The OS/hypervisor for any PCIe device loads an appropriate driver.

Depending on the system configuration stored in flash and dynamically modified by UCS Manager (see Section 6.1.), the embedded management CPU can create any legal PCIe topology and device combination and make it visible to the host. Palo is capable of a maximum of 64K PCIe devices, even if there are only 128 useful endpoints supported.

Palo is a data center class adapter, all internal and external memories for data and control are ECC protected, and it has diagnostics LEDs. As we have already discussed, it fully supports the DCB and FCoE standards and it support virtualization through vNIC, PCIe topologies and VNTag.

Focusing on the Ethernet vNIC aspect, Palo supports interrupt coalescing (MSI and MSI-X), Receive Side Scaling (RSS), stateless offload acceleration (checksum, LSO: Large Segment Offload), IEEE 802.1Q VLAN trunking, IEEE 802.1p priorities, jumbo frames (9KB), 32 multicast filters, promiscuous mode, and PXE.

The FC vNIC is depicted in Figure 87 that shows that the SCSI protocol is supported in HW by the eCPU (embedded CPU), but a path is also provided for sending raw FC frames (mainly for control and management protocols). Also data sequencing and FC frame generation is handled by dedicated hardware engines

Palo supports 2048 concurrent logins and 8,192 active exchanges; it is capable to operate as a SCSI initiator or target; it supports VSANs (Virtual SANs) and NPIV (N_Port Id Virtualization).

Some preliminary performance data are reported in the following.

At the hardware layer it is possible to saturate both ports in both directions, but this is not a particularly interesting number, since the performance perceived by the user is at the application layer.

Using Linux netperf with 1,500 bytes MTU, enabling a single port, the throughput measured was 9.4 Gb/s measured at the application layer (basically 10 Gb/s minus the overhead of the headers). This holds for both simplex

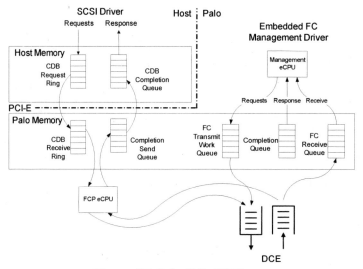

Figure 87: Palo FC vNIC

and full duplex traffic.

For FC, the preliminary performance measured is as follow:

- 512 byte reads: 590K IOPS
- 512 byte write: 500K IOPS

In term of latency, using two hosts with Palo connected back to back, Linux NetPIPE latency is 9.8 usec (microseconds), measured at the application latency. The hardware latency of the two Palo connected back-to-back is 3 usec, i.e., approximately 2 usec for the sending Palo and 1 usec for the receiving Palo.

Palo is supported on all OSes and hypervisors that are supported in California.

4.7. Emulex

Emulex[5] has a new family of CNAs that leverages the Ethernet enhancements and Fibre Channel over Ethernet (FCoE) technology to drive convergence over the underlying 10G Enhanced Ethernet infrastructure.

To the operating systems accessing the CNA, the adapter's functionality is near transparent, since it independently presents both the NIC and the HBA functionality (see Figure 88 and Figure 90).

While different traffic types share a single Ethernet link, the CNA has the functionality to provision guaranteed bandwidth for each traffic type through the use of traffic groups. This capability to configure bandwidth to different traffic groups enables IT administrators to support "wire once" mode of deployment and configure the bandwidth dynamically as the work load of the applications changes on a particular server.

The core functionality of the CNA is that it offloads Fibre Channel protocol processing from the CPU thus providing high performance storage connectivity while improving the overall CPU utilization of the server. The additional processing cycles thus saved by the CPUs provide higher CPU bandwidth for enterprise applications. In server virtualization environments, this additional CPU bandwidth enables deployment of more virtual machines per server.

---

5   The authors are thankful to Emulex® Corporation for the information and the material provided to edit this section. Pictures are courtesy of Emulex®.

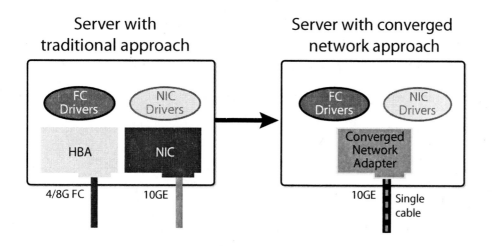

Figure 88: CNAs drivers

4.7.1. Emulex® LP2105-CI

LP2105-CI is the Emulex® multi-protocol CNA made available in mezzanine card format to suit the requirements of the California system (the mezzanine card is named Cisco UCS CNA M71KR-E – Emulex Converged Network Adapter), see Figure 89.

The LP2105-CI consists of a 10Gb/s Ethernet Controller, a 4Gb/s Fibre Channel controller and an Enhanced Ethernet bridging chipset that provides the enhanced Ethernet functionality in addition to encapsulation of Fibre Channel frames on to Ethernet frames. In order to support transmission of Fibre Channel data frames without any fragmentation and defragmentation, the 2KB Fibre Channel frames are mapped on to mini-jumbo frames of 2.5KB thus ensuring that there no additional latencies introduced in the processing of storage traffic.

The adapter connects to the host bus through two PCIe interfaces – one for the 10Gb/s Ethernet controller (8 PCIe lanes) and the other for the Fibre Channel controller (4 PCIe lanes). The Fibre Channel controller leverages seven generations field-proven Emulex® LightPulse® technology to deliver a distinctive advantage with respect to seamless interoperability with existing FC storage infrastructure.

Figure 91 shows the Emulex® HBAnyware® application that provides functions for managing the converged network adapter as well as for assigning bandwidth for individual traffic types.

Figure 89: UCS CNA M71KR-E

The Enhanced Ethernet bridging chipset provides the capability to isolate and prioritize different traffic types on the link and by supporting bandwidth guarantees for different traffic types, thus enabling the CNA to support efficient consolidation of Fibre Channel and Ethernet traffic on to a single link without any degradation in performance.

Key Features of LP2105-CI:

- Two 10Gb Enhanced Ethernet ports (supporting FCoE);
- Host Interface: PCIe v1.1:
  - PCIe x4 interface to Fibre Channel;
  - PCIe x8 interface to Ethernet Controller;
- Mezzanine Card's back plane interconnect: 10G-Base-KR interface;
- Mezzanine form factor of 3.65" x 7.25".

4.7.2. Fibre Channel Features

The LP2105-CI has the following fibre channel features:

- Offloads Fibre Channel protocol processing to the FC controller;
- Exceptional performance and full-duplex data throughput;
- Comprehensive virtualization capabilities with support for N-Port

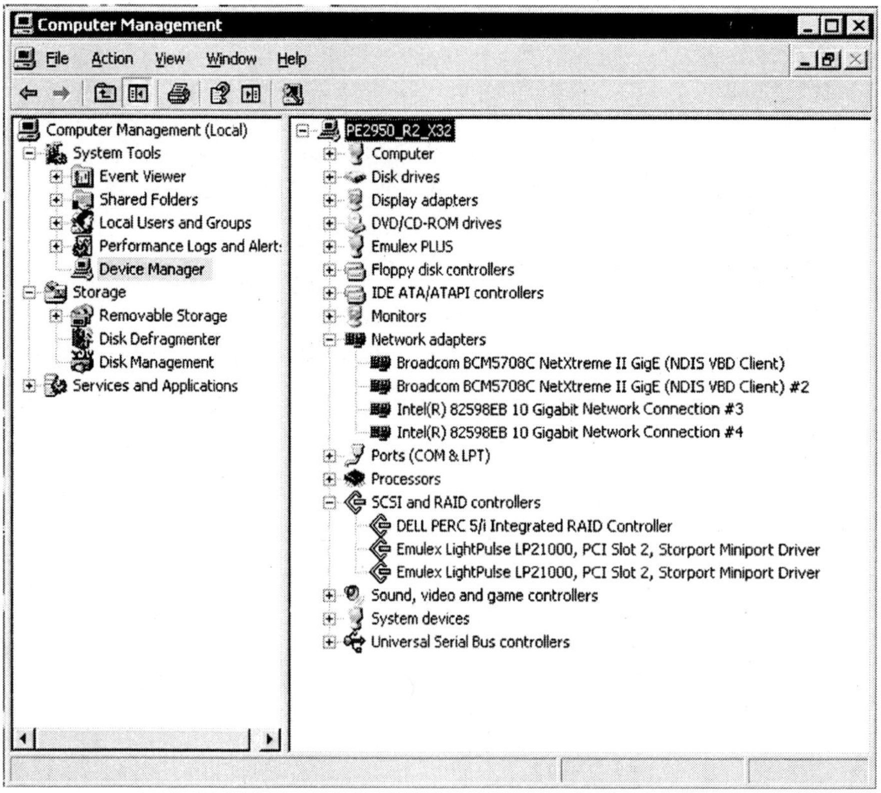

Figure 90: CNA drivers for Windows

ID Virtualization (NPIV) and Virtual Fabric;

- Simplified installation and configuration;

- Efficient administration via Emulex® HBAnyware;

- Strong authentication between host and fabric based on Fibre Channel Security Protocol (FC-SP) Diffie-Helmann Challenge Handshake Authentication Protocol (DH-CHAP).

4.7.3. Ethernet features

The LP2105-CI has the following Ethernet features:

- Mini-jumbo frames 2.5Kbytes for avoiding fragmentation delays;

- VLAN Tagging (802.1Q);

- Flow Control based on 802.3x;

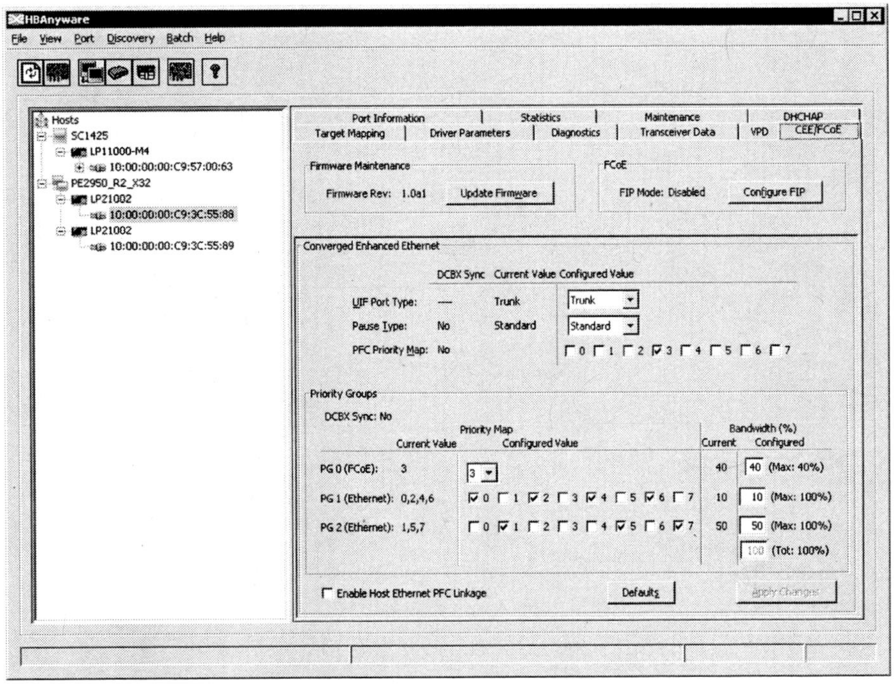

Figure 91: Emulex® HBAnyware®

- Congestion Management;

- Per Priority Pause / Priority Flow Control;

- Advanced Packet Filtering;

- Tx TCP Segmentation offload;

- IP, TCP, UDP Checksum offload capabilities.

4.7.4. Functional Architecture

The Emulex® LP2105-CI provides Converged Network Adapter functionality while appearing to the host as two PCI devices—a network adapter and Fibre Channel adapter. Host networking and storage drivers communicate with the appropriate PCI function in the Emulex® LP2105-CI.

The OS actually sees a 2-port 10GE adapter and a 2-port SCSI FC card, see Figure 90.

I/O operations are processed by the respective functionality of the CNA (see Figure 92), and in the case of a networking transaction, delivered to the lossless Media Access Control (MAC) for delivery to the unified fabric. For Fibre

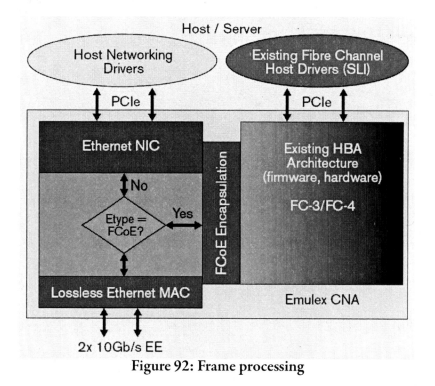

Figure 92: Frame processing

Channel transactions, Fibre Channel frames are sent to the FCoE encapsulation engine, and then transferred to the lossless MAC for delivery to the unified fabric.

Received traffic is handled much the same way. Incoming traffic is processed by the lossless MAC, and then filtered based on FCoE Ethertype. Non-FCoE traffic delivered to the Ethernet NIC, and FCoE traffic de-capsulated by the FCoE engine, are then forwarded to the Fibre Channel HBA device for further processing. The block is then relayed to the appropriate host device driver.

4.7.5. Deployment in Project California

The Emulex® LP2105-CI consolidates multiple traffic types on to a single 10 Gb/s enhanced Ethernet link.

The Emulex® LP2105-CI connects to the mid-plane using the 10G-Base-KR interfaces and thus provides the connectivity to the California fabric (see Figure 93).

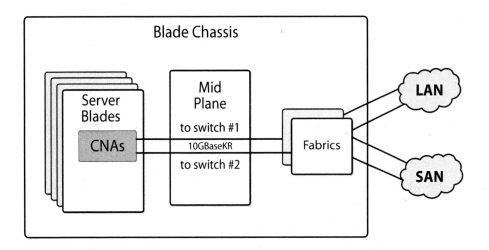

Figure 93: Emulex® CNA inside California

4.7.6. *Management of CNAs*

Reducing the time it takes for installing and managing an infrastructure is a key objective for IT administrators and the goal is to keep the time spent on these tasks to a minimum. Fibre Channel technology has helped administrators maintain a highly reliable and secure infrastructure for the most critical enterprise data.

From installation to management, Emulex® provides an extensive collection of management software and utilities to simplify the deployment and manageability of the CNAs across the data center. The installation utility streamlines the deployment of Emulex® CNA drivers. The scripting capabilities supported on the utility enable the driver installation process to be automated for SAN wide deployment. The Emulex® HBAnyware management suite delivers a powerful management application. The use of industry standard HBA APIs for management facilitates easy upward integration with element management system (EMS) applications. This level of integration greatly simplifies the process of provisioning and deploying new servers with converged network connectivity, thereby saving IT departments both time and money.

The Emulex® HBAnyware provides the functionality wherein Fibre Channel HBAs and FCoE functionality of the CNAs can be managed through the same application (see Figure 94) – thus making it easier for IT managers to deploy Converged Network Adapters without any impact on their existing processes.

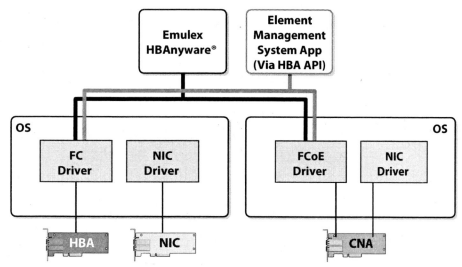

Figure 94: HBAs and CNA management

4.7.7. Benefits of Emulex® selection

By installing an Emulex® CNA in California, a customers has the following advantages:

- Emulex® CNAs provide high performing, reliable storage connectivity by leveraging the field proven Emulex® LightPulse architecture in the CNAs. The adapters also make use of common driver model across FCoE CNAs and Fibre Channel HBAs enabling simpler driver management and better scalability.

- The Emulex® CNAs host industry proven FC functionality that implies extensive interoperability with existing Fibre Channel investments such as switches, directors and Fibre Channel storage systems.

- The FCoE drivers used for the CNAs are qualified and supported by major operating system such as Windows, Linux, VMware and Solaris with additional platforms in the roadmap.

- Emulex's management suite, HBAnyware, is highly intuitive and provides a single window for managing the newly deployed CNAs as well as existing Fibre Channel adapters.

- Emulex® CNAs are certified to work with all the major multipathing solutions in the market today from server, storage and operating system vendors.

- Emulex® CNAs support Boot from SAN which enable efficient management of OS images and server migration in the data center.

4.8. QLogic®

QLogic® Corporation is a long-time market leader in SAN Fibre Channel adapters and iSCSI / Ethernet adapters for both standard discrete servers and the mezzanine form-factor for blade servers. Since 2004, QLogic® has held the number one market share position for SAN adapters.[6]

QLogic® provides the ASIC for a Cisco designed and manufactured mezzanine card for California named Cisco UCS CNA M71KR-Q - QLogic Converged Network Adapter, which will be followed by other models.

The ideal CNA needs to combine all the features, capabilities, and performance of a best-in-class Ethernet NIC and best-in-class Fibre Channel SAN host bus adapter into a highly-reliable, low-power, cost-effective converged network adapter. However, in order for the benefits to be fully realized, neither the Ethernet NIC functionality nor the FC HBA functionality can be substandard or compromised in any way. Customers have grown accustomed to OEM-hardened LAN / SAN software stacks, broad operating system driver support, and interoperability, which are now baseline requirements for any enterprise operation.

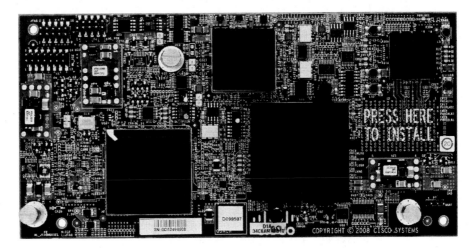

Figure 95: UCS CNA M71KR-Q

---

6   The authors are thankful to QLogic® Corporation for the information and the material provided to write this section.

QLogic's family of converged network adapters draws upon the intellectual property of two successful families of QLogic® products. One is the highly successful iSCSI/Ethernet product family, in use today by multiple OEM server providers in both discrete and blade server form factor. The other is the market share leading Fibre Channel SAN adapter family, in use today by every server and storage OEM server provider, in both discrete and blade server form factor. Leveraging the architecture, features, capabilities, performance, and software maturity of these product families makes the QLogic® CNA the best in class, from day one. QLogic's leadership position in adapters translates directly into a better CNA product.

4.8.1. High Performance

QLogic® CNAs boost system performance with 10Gbps speed and full hardware offload for FCoE protocol processing. Leading edge 10Gbps bandwidth eliminates performance bottlenecks in the I/O path with a 10X data rate improvement versus existing 1Gbps Ethernet solutions. Additionally, full hardware offload for FCoE protocol processing reduces system CPU utilization for I/O operations, which leads to faster application performance and higher levels of consolidation in virtualized systems:

- 10Gbps per port maximum throughput for high bandwidth storage (SAN) and networking (LAN) traffic;

- Full hardware offload for FCoE protocol processing;

- Over 150,000 IOPS per port deliver high I/O transfer rates for storage applications;

- Full support for TCP/IP and Ethernet performance enhancements such as priority-based flow control (802.1 Qbb), jumbo frames, checksum offloads, and segmentation offloads;

- PCIe 2.0 (Gen2) bus interface.

4.8.2. Investment Protection

QLogic® CNAs are designed to preserve existing investment in Fibre Channel storage and core Ethernet switches and routers for data networking. They leverage the same software and driver stacks that have been deployed and battle-hardened in millions of previous installations.

- Works seamlessly with existing Fibre Channel storage;

- Communicates via fully-featured enhanced Ethernet, the most common networking technology in the world;

- Compatible with existing Fibre Channel drivers that have been deployed in millions of existing systems;

- Broad operating system driver support;

- Standards compliant.

4.8.3. Lower Total Cost of Ownership (TCO)

QLogic® CNAs reduce data center costs through convergence. Now, one CNA can do the work of a discrete FC Host Bus Adapter (HBA) and Ethernet NIC. This convergence also means fewer cables, fewer switches, less power consumption, reduced cooling, and easier LAN and SAN management. They preserve familiar FC concepts such as WWNs, FC-IDs, LUN masking, and zoning, thereby eliminating training costs that would be required for a new storage technology.

- Reduced hardware, cabling, power, cooling, and management costs through convergence of data and storage networking;

- Preservation of familiar data and storage concepts resulting in lower training and administrative costs.

5. Cisco® Unified Computing System

The Cisco® Unified Computing System (UCS), aka Project California or simply California is a data center server that unifies network virtualization, storage virtualization, and server virtualization, within open industry standard technologies and with the network as the platform.

This chapter will first give a brief overview of the various components and then describe them in detail.

5.1. Components overview

Figure 96 shows the components of a California system, with the exception of the UCS manager which is a software element.

5.1.1. UCS Manager

The Cisco® UCS Manager software integrates the components of a Cisco® Unified Computing System into a single, seamless entity. It manages all the server blades of a UCS as a single logical domain using an intuitive GUI with both

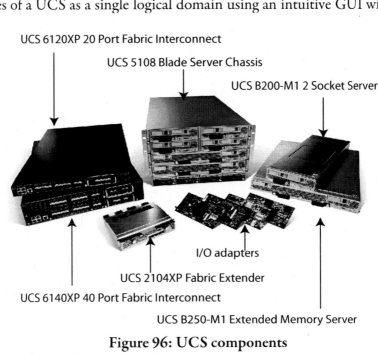

Figure 96: UCS components

CLI and XML API options, enabling near real time configuration and re-configuration of resources (see Section 6.). Tasks that once required multiple administrators (server, network, and storage) and days or hours now can be accomplished in minutes, with reduced risk for human errors. The software's role-based design supports existing best practices, allowing server, network, and storage administrators to contribute their specific subject matter expertise to a system design. Any user's role may be limited to a subset of the system's resources using organizations and locales, so that a Cisco® Unified Computing System can be partitioned and shared between organizations using a multi-tenant model. For organizations implementing Information Technology Infrastructure Library (ITIL) based processes, the Cisco® UCS Manager helps to codify and enforce best practices. It maintains and conducts all operations based on an internal configuration database that can be exported to populate CMDBs (Configuration Management Data Bases) and integrate with higher-level software provisioning tools.

5.1.2.   UCS 6100 Series Fabric Interconnects

Cisco's UCS 6100 Series Fabric Interconnects, a family of line-rate, low-latency, lossless 10 Gigabit Ethernet, Cisco® Data Center Ethernet, and Fiber Channel over Ethernet (FCoE) switches, consolidate I/O at the system level. Leveraging upon the same switching technology as the Cisco® Nexus™ 5000 Series, the Cisco® Unified Computing System Series 6100 Fabric Interconnects provide the additional features and management capabilities that make up the central nervous system of the Cisco® Unified Computing System. The Fabric Interconnects provide a unified network fabric that connects every server resource in the system via wire once 10GE/FCoE downlinks and flexible 10GE and 1/2/4 GFC uplink modules. Out of band management, switch redundancy, and console-based diagnostics are enabled through dedicated management, clustering, and RS-232 ports. A single UCS 6100 Series Fabric Interconnect unites up to 320 servers within a single system domain for large scalability. The switches feature front to back cooling, redundant front hot-pluggable fans and power supplies, and rear cabling enabling efficient cooling and serviceability. Typically deployed in active-active redundant pairs, Fabric Interconnects provide uniform access to both networks and storage, eliminating the barriers to deploying a fully virtualized environment based on a flexible, programmable pool of resources.

UCS 6120XP 20 Port Fabric Interconnect

The Cisco® UCS U6120XP Fabric Interconnect is 1 RU height, featuring twenty 10GE/FCoE ports. Additional ports are available via an Expansion Module. At the time of writing the following uplink modules exist:

- 6 ports 10 GE with DCE/FCoE support;

- 4 ports 10 GE with DCE/FCoE support and 4 ports Fibre Channel 1/2/4 Gbps;

- 8 ports Fibre Channel 1/2/4 Gbps;

- 4 ports Fibre Channel 2/4/8 Gbps (not supported at first customer shipment).

With Cisco's UCS Manager embedded, this Fabric Interconnect can support up to 160 blade servers/20 chassis in a single domain.

UCS 6140XP 40 Port Fabric Interconnect

The Cisco® UCS U6140XP Fabric Interconnect is 2 RU height, featuring 40 10GE/FCoE ports. Additional expansion is available via two Expansion Modules (the same used in the 6120XP) . With Cisco's UCS Manager embedded, this Fabric Interconnect can support up to 320 servers/40 chassis in a single domain.

5.1.3. UCS 2100 Series Fabric Extenders

At the time of writing there is a single fabric extender, the UCS 2104XP.

UCS 2104XP Fabric Extender

The Cisco® UCS U2104XP Fabric Extender extends the I/O fabric into the blade server enclosure providing one to four 10GE connections between the blade enclosure and the Fabric Interconnect, simplifying diagnostics, cabling and management.

The Fabric Extender multiplexes and forwards all traffic using a cut-through architecture to the Fabric Interconnect where the Fabric Interconnect manages network profiles efficiently and effectively. Each Fabric Extender has eight 10GBASE-KR connections to the blade enclosure mid-plane, one connection per each half slot. This gives each half-slot blade server access to two 10GE unified-fabric links offering maximum throughput and redundancy. The Cisco® UCS U2104XP Fabric Extender features an integrated chassis management controller for the blade enclosure physical components, including power supplies, fans, and temperature sensors. The fabric extender also connects to each blade's management port for management, monitoring, and firmware updates.

5.1.4. UCS 5100 Series Blade Server Chassis

At the time of writing there is a single blade server chassis, the UCS 5108, shown in Figure 97.

UCS 5108 Blade Server Chassis

The Cisco® UCS U5108 Blade Server Enclosure physically houses blade servers and up to two fabric extenders. The enclosure is 6RU high, allowing for up to 7 enclosures and a total of 56 servers per rack. Compared to complex traditional blade enclosures, the U5108 Blade Server Enclosure is dramatically simple in its design. It supports up to eight half slot or four full slot blade servers with four power supplies, and eight cooling fans. Both power supplies and fans are redundant and hot swappable. Featuring 90%+ efficient power supplies, front to rear cooling, and airflow optimized mid-plane, the Cisco® UCS U5108 is optimized for energy efficiency and reliability.

Figure 97: UCS 5108

5.1.5. UCS B-Series Blade Servers

The Cisco® UCS B-Series Blade Servers are designed for compatibility, performance, energy efficiency, large memory footprints, manageability, and unified I/O connectivity. Based on Intel® Xeon® 5500 series processors, B-Series Blade Servers adapt to application demands, intelligently scale energy use, and offer best in class virtualization. Each Cisco® UCS B-Series Blade Server utilizes converged network adapters for consolidated access to the unified fabric with various levels of transparency to the operating system. This design reduces the number of adapters, cables, and access-layer switches for LAN and SAN connectivity at the rack level. This Cisco® innovation significantly reduces capital and operational expenses, including administrative overhead, power, and cooling.

UCS B200-M1 2 Socket Server

The Cisco® UCS B200-M1 Two Socket Blade Server is a half-slot, two-socket blade server (Figure 96 shows it with the metal sheet, Figure 98 shows just the board). The system features two Intel® Xeon® 5500 Series processors, up to 96GB of DDR3 memory, two optional small form factor SAS disk drives, and a dual-port converged network adapter mezzanine slot for up to 20Gbps of I/O throughput. The B200-M1 balances simplicity, performance, and density for mainstream virtualization and other datacenter workload performance.

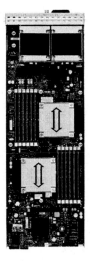

Figure 98: UCS B200-M1 2 Socket Server

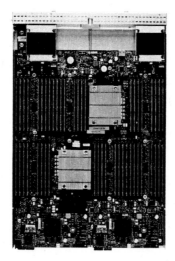

Figure 99: UCS B250-M1 Extended Memory Server

UCS B250-M1 Extended Memory Server

The Cisco® UCS B250-M1 Two Socket Blade Server is a full-slot, two-socket blade server featuring Cisco® Extended Memory Technology (Figure 96 shows it with the metal sheet, Figure 99 shows just the board). The system features two Intel® Xeon® 5500 Series processors, up to 384GB of DDR3 memory, two optional small form factor SAS disk drives, and two dual-port converged network adapter mezzanine slots for up to 40Gbps of I/O throughput. The B250-M1 maximizes performance and capacity for demanding virtualization and memory intensive workloads, with greater memory capacity and throughput.

5.1.6. I/O adapters

Four I/O adapters are available at the time of writing:

- Cisco® UCS VIC M81KR – Intel® Virtual Interface Card (see Section 4.6.);

- Cisco® UCS CNA M71KR-E – Emulex® Converged Network Adapter (see Section 4.6.);

- Cisco® UCS CNA M71KR-Q – QLogic® Converged Network Adapter (see Section 4.8.);

- Cisco® UCS 82598KR-CI – Cisco® Converged Network Adapter (see Section 4.3.).

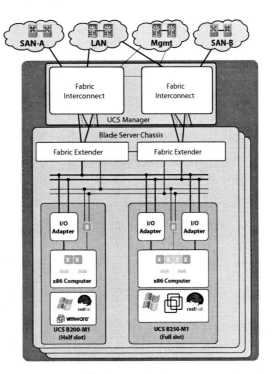

Figure 100: Component interconnection

5.1.7. Overall organization

A California system is organized as shown in Figure 100. There is one or typically two fabric interconnects that have uplinks to the LAN, SANs (two separate SANs are supported) and to the management network. The 10GE downlinks connect to the blade server chassis: any number of chassis from 1 to 40 is supported. Each blade server chassis contains one or typically two fabric extenders, the fabric extenders connect to the I/O adapters and to the management processors on each server blade (the connection are point to point, but are shown in Figure 100 as busses for graphic simplicity).

A UCS 5108 Blade Server Chassis has eight half-slots. Each pair of half-slots can be converted into a full slot by removing a metal sheet separator. This allow the installation of either:

- 8 half-slot server blades, or
- 6 half-slot server blades and 1 full-slot server blade, or
- 4 half-slot server blades and 2 full-slot server blades, or
- 2 half-slot server blades and 3 full-slot server blades, or
- 4 full-slot server blades.

5.2.   Detailed Description

The next few sections will add detail to the previous descriptions.

5.2.1.   UCS 6100 Series Fabric Interconnects

The UCS 6100 Fabric Interconnects are depicted in Figure 101. On top is the 6120XP and on the bottom the 6140XP. The 6120XP has twenty 10GE ports that support DCE (Data Center Ethernet, see Section 3.1.1.) and FCoE (see Section 3.1.7.), and one Expansion module. The 6140XP has double the ports and performance of the 6120XP. Detailed physical data can be found in Section 7.2.

Different Expansion modules are available to provide additional Ethernet and Fibre Channel connectivity. All the Ethernet ports (not only those in the expansion module) can be used as uplinks toward a network backbone or as downlinks toward the Fabric Extenders. The Fibre Channel ports can only be used as uplink toward Fibre Channel fabrics.

Ethernet connectivity

In the definition of UCS (see Section 1.3.) it became clear that the system is network centric. As such it can be configured to behave on the Ethernet up-links either as a group of hosts or as a switch.

Figure 101: UCS 6120XP & UCS 6140XP

The UCS 6100 Fabric Interconnects are derived from the hardware of the Nexus 5000 and as such they can support a variety of network protocols and are capable of acting as Ethernet switches. To explain how Ethernet connectivity works refer to Figure 102.

FI1 and FI2 are two UCS 6100 (the number of ports does not matter) and SW1 and SW2 are two data center switches like the Nexus 5000 or the Nexus 7000.

Spanning Tree

The first possibility is to configure FI1 and FI2 as Ethernet switches. In this case the spanning tree protocol needs to be enabled and it detects a loop FI1 – A – SW1 – B – C – SW2– D– FI1. Spanning tree needs to prune any loop into a tree and probably it blocks either port A or port D (the root of the spanning tree is typically either in SW1 or SW2). This is a suboptimal solution because, even if it provides high availability, it does not use all the uplinks, therefore wasting expensive bandwidth.

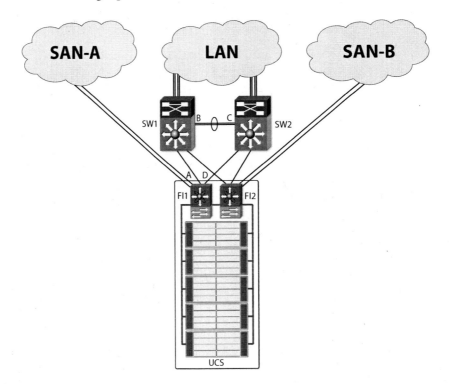

Figure 102: UCS External connectivity

To solve this issue several different techniques can be used.

EtherChannel

The first possibility is to connect FI1 to a single switch (e.g. SW1) thus avoiding any loop. The connection itself can be made highly available using multiple links in parallel with EtherChannel.

EtherChannel is a port aggregation technology primarily introduced by Cisco® in 1994 and standardized by IEEE in 2000 in the project IEEE 802.3ad. EtherChannel allows aggregating several physical Ethernet links to create one logical Ethernet link with a bandwidth equal to the sum of the bandwidths of the aggregated links. EtherChannel can aggregate from 2 to 16 links and all higher-level protocols see these multiple links as a single connection. This is beneficial for providing fault-tolerance and high-speed links between Fabric Interconnects and the LAN backbone, without blocking any port and therefore using all the links. A limitation of EtherChannel is that all the physical ports in the aggregation group must be between one Fabric Interconnect and one LAN switch. For this reason the next three solutions – VSS, vPC and Ethernet Host Virtualizer – were developed.

VSS

VSS (Virtual Switching System) is the first of two Cisco® technologies that allows using EtherChannel from a Fabric Interconnect to two LAN switches avoiding any blocked port. VSS accomplishes this by clustering the two LAN switches SW1 and SW2 into a single, managed, and logical entity. The individual chassis become indistinguishable and therefore each Fabric Interconnect believes the upstream LAN switches to be a single STP (Spanning Tree Protocol) bridge, and EtherChannel can be deployed unmodified on the Fabric Interconnect. VSS has also other advantages, since it improves high availability, scalability, management, and maintenance, see [17] for more details. Today VSS is deployed on the Catalyst 6500 switches.

vPC

vPC (virtual Port Channel), aka MCEC (Multi-Chassis EtherChannel), achieves a result similar to VSS without requiring the clustering of the two LAN switches.

From the Fabric interconnect perspective nothing changes: it continues to use an unmodified EtherChannel and it sees the vPC switches as a single LAN switch.

The two LAN switches coordinate who is in charge of delivering each frame to the Fabric Interconnect, fully utilizing both links. The key challenge in vPC (common also to VSS) is to deliver each frame exactly once, avoiding frame duplication and loops. This must also be guaranteed when some Fabric Interconnect that are connected to both LAN switches and some that are connected to only one (because, for example, one uplink has failed), without using Spanning Tree.

Ethernet Host Virtualizer

VSS and vPC are techniques implemented on the LAN switches to allow the Fabric Interconnects to keep using EtherChannel in a traditional manner.

In addition, the same problem can also be solved on the Fabric Interconnect by a technique called Ethernet Host Virtualizer (aka End Host Mode), that is the preferred technology choice.

With reference to Figure 102, the Fabric Interconnect implements Ethernet Host Virtualizer while the LAN switches continue to run the classical Spanning Tree Protocol.

A Fabric Interconnect running Ethernet Host Virtualizer divides its ports into two groups: host ports and network ports. Both types of ports can be a single interface or an EtherChannel. The switch then associates each host port with a network port. This process is called "pinning". The same host port always uses the same network port, unless it fails. In this case, the Fabric Interconnect moves the pinning to another network port (dynamic pinning).

In the example of Figure 103 , MAC-A is always presented on the left network port and MAC-B is always presented on the right network port.

For a given host port, deciding which network port to use may be either based on manual configuration or the Fabric Interconnect (according to the load) may determine the network port. The relationship remains intact until either the host port or the network port loses connectivity.

When this happens, the associated host ports are redistributed to the remaining set of active network ports. Particular attention must be paid to multicast and broadcast frames to avoid loops and frame duplications. Typically Fabric Interconnects act as follows:

- They never retransmit a frame received from a network port to an-

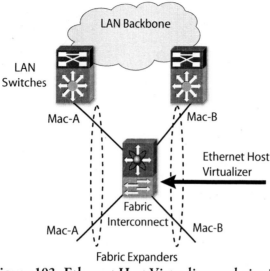

Figure 103: Ethernet Host Virtualizer and pinning

other network port;

- They divide the multicast/broadcast traffic according to multicast groups and they assign each multicast group to a single network port. Only one network port may transmit and receive a given multicast group.

The Ethernet Host Virtualizer is the preferred configuration at UCS first customer shipment since it can plug in any LAN network, without requiring any modification.

Fibre Channel Connectivity

The Fibre Channel connectivity has historically used a different high availability mode than the Ethernet connectivity. Most Fibre Channel Installations use two separate SANs (normally called SAN-A and SAN-B) built with two different sets of Fibre Channel switches. Each host and storage array connects to both SANs using two separate HBAs. High Availability is achieved at the application level by running a multipathing software that balances the traffic across the two SANs using either an active/active or active/standby mode.

A California system supports this model by having two separate Fabric Interconnects, two separate Fabric Extenders and dual-port CNAs on the blades. One Fabric Interconnect with all the Fabric Extenders connected to it belongs to SAN-A and the other Fabric Interconnect with all the Fabric Extenders connected to it belongs to SAN-B. The two SANs are kept fully separated as in the classical Fibre Channel model.

As in the case of Ethernet, each Fabric Interconnect is capable of presenting itself as a FC switch or as a FC host (NPV mode).

Fabric Interconnect as FC switch

This is a theoretical possibility that the hardware supports and it is standards compliant. It consists of running the FC switching software on the Fabric Interconnect and using E_Ports (Inter Switch Links) to connect to the FC backbones. With reference to Figure 102, the links between FI1 and SAN-A and FI2 and SAN-B are E_Port on both ends.

Unfortunately this implies assigning a FC domain_ID to each California and since the number of domain_IDs is typically limited to 64, it is not a scalable solution. Some storage manufacturers support a number of domain that is much smaller than 64 and this further limits the applicability of this solution.

Fabric Interconnect as a host

This solution is based on a concept that has recently been added to the FC standard and it is know as NPIV (N_Port ID Virtualization) a Fibre Channel facility allowing multiple N_Port IDs (aka FC_IDs) to share a single physical N_Port. The term NPIV is used when this feature is implemented on the host, for example to allow multiple virtual machines to share the same FC connection. The term NPV (N_Port Virtualization) is used when this feature is implemented in an external switch that aggregates multiple N_Ports into one or more uplinks. A NPV box behaves as an NPIV-based HBA to the core Fibre Channel switches. According to these definitions each Fabric Interconnect can be configured in NPV mode, i.e.:

- Each Fabric Interconnect presents itself to the FC network as a host, i.e. it uses an N_Port (Node Port);

- The N_Port on the Fabric Interconnect is connected to an F_Port (Fabric Port) on the Fibre Channel Network;

- The Fabric Interconnect performs the first FLOGI to bring-up the link between the N_Port and the F_Port;

- The FLOGIs received by the Fabric Interconnect from the server's adapter are translated to FDISCs according to the NPIV standard.

This eliminates the scalability issue, since it does not assign a FC domain_ID to each Fabric Interconnect. It also greatly simplifies interoperability, since multivendor interoperability is much better in FC between N_Port and F_Port as opposed to E_Port and E_Port. Finally, it guarantees the same high availability present today in a pure FC installation by fully preserving the dual fabric model.

With reference to Figure 102 the links between FI1 and SAN-A and FI2 and SAN-B are N_Ports on the Fabric Interconnect side and F_Ports on the SAN side.

Value added features that can be used in NPV mode are F_Port Trunking and F_Port Channeling.

F_Port Channeling is similar to EtherChannel but it applies to FC. It is the bundling of multiple physical interfaces into one logical high-bandwidth link. F_Port Channeling provides higher bandwidth, increased link redundancy, and load balancing between a Fabric Interconnect and a FC switch.

F_Port Trunking allows a single F_Port to carry the traffic of multiple VSANs, according to the FC standards.

5.2.2. UCS 2104XP Fabric Extender

The Cisco® UCS 2104XP Fabric Extender (aka FEX, see Figure 104) is a special blade that plugs in the rear of a UCS 5108 chassis (see Figure 107).

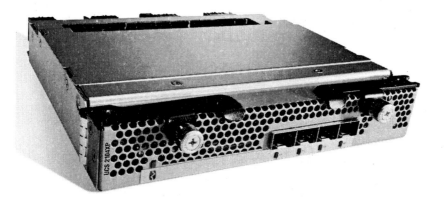

Figure 104: UCS 2104XP Fabric Extender

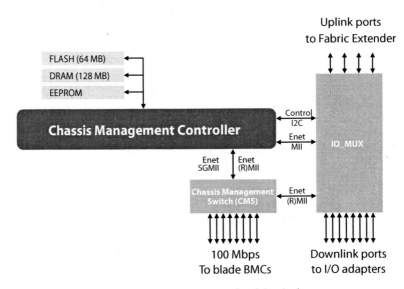

Figure 105: Fabric Extender block diagram

Its block diagram is shown in Figure 105. The UCS 2104XP Fabric Extender has three subsystems inside:

- The Redwood IO_MUX;
- The Chassis Management Controller (CMC);
- The Chassis Management Switch (CMS).

Each of these subsystems is designed to perform a specific task in the California system.

Redwood IO_MUX

The Redwood IO_MUX is used as a bridge between the server blades and the Fabric Interconnect. It is the ASIC that implements the data plane of the Fabric Extender. It provides:

- Eight 10GE external downlink ports to connect the server blades;
- Four 10GE external uplink ports to connect to the fabric interconnect;
- One 1GE internal port to connect the CMS;
- One 100Mbps internal port towards the CMC.

By default mezzanine adapters installed on server blades get pinned to uplinks in a pre-determined fashion. This method makes it easy to understand which blade uses which uplink. Each half-slot supports one adapter. The UCS B200-M1 blade fits into a half-slot and therefore supports a single adapter, but the UCS B250-M1 takes a full-slot and can have two adapters.

Future blade offerings could even increase the number of slots that a single blade can use. Table 7 gives an understanding of the server pinning to the uplinks in different scenarios.

In future firmware releases the configuration of slot pinning to an uplink will be a user configurable feature.

Number of Links from Fabric Extender to Fabric Interconnect = 1	
Uplink	Slots pinned to uplink
1	1,2,3,4,5,6,7,8
Number of Links from Fabric Extender to Fabric Interconnect = 2	
Uplink	Slots pinned to uplink
1	1,3,5,7
2	2,4,6,8
Number of Links from Fabric Extender to Fabric Interconnect = 4	
Uplink	Slots pinned to uplink
1	1,5
2	2,6
3	3,7
4	4,8

Table 7: Uplink pinning

The uplinks from the Fabric Extender (i.e., Redwood) to the Fabric Interconnect use the VNTag header (see Section 3.2.6.). A particularly important field inside the VNTag is the VIF (Virtual InterFace) that is used to identify the downlink port (ports in the case of a broadcast/multicast frame).

VNTag may also be used between the adapter and the Fabric Extender on the downlink ports, but this requires an adapter that is VNTag capable, like the one based on Menlo (see Section 4.7. and Section 4.8.) or Palo (see Section 4.6.).

Frame Flow in Redwood – Host to Network

With reference to Figure 105:

- The frame arrives at the downlink port from the mezzanine card;
- The destination uplink port is selected as a function off the port on which the frame was received;
- The frame is stored in a receiver buffer and queued to an uplink port (virtual output queuing);
  - Virtual output queues are per output, per CoS (Class of Service);
- The appropriate VNTag is added to the frame;
- The frame is transmitted on the selected Uplink port;
  - Frame transmit arbitration is per destination Uplink port; it is based on a round-robin algorithm for CoS and in order delivery inside each CoS;
- Any given frame is forwarded on a single uplink.

Frame Flow in Redwood – Network to Host

With reference to Figure 105:

- The frame arrives at the uplink port from the Fabric Interconnect;
- The destination downlink port is selected according to the VNTag header;
  - The decision is based on the destination VIF;
- The frame is stored in the receiver buffer and queued at the downlink port;
  - Separate resources are available for different CoS values;
- A single frame can be forwarded to multiple destinations (e.g., broadcast/multicast frames).

It is important to understand that there is no local switching function in the Redwood ASIC. A frame coming from a downlink port cannot be forwarded to another downlink port. The multicast/broadcast replication is done only from an uplink port toward downlink ports, not vice-versa.

Chassis Management Controller – CMC

The Chassis Management Controller (CMC) is a processor embedded in the Fabric Extender. The CMC interacts with the UCS Manager and the Baseboard Management Controllers (BMCs) present on the server blades. The administrator does not interact directly with the CMC, but only through the UCSM. The CMC main function is to provide overall chassis discovery and management and to report the result to the UCS manager. It also provides platform services for the Redwood management software.

The CMC implements seven main functions:

- It controls the chassis fans;
- It monitors and logs fan speed;
- It monitors and logs ingress and egress temperatures;
- It controls location indication and chassis fault indications;
- It powers up/powers down power supplies, monitoring and logging voltages, currents and temperatures inside the chassis;
- It detects presence, insertion and removal of UCS blades;
- It reads the IDs of the chassis, UCS blades, and Fabric Extenders.

It is important to understand that CMC does not manage UCS blades.

If two UCS2104XP Fabric Extenders are installed in a chassis, the two CMC processors automatically form a cluster and only one of them will be active at a given time. A high availability algorithm between the two CMCs defines the active CMC. A serial interface is used for heartbeats between the two CMCs. Failover is triggered either by loss of heartbeat, or if the active CMC is unplugged for any reason. The UCS Manager can also force a fail-over. The information about the active CMC is stored into a SEEPROM (Serial EPROM) and the CMC must read it before accessing shared resources.

Chassis Management Switch (CMS)

The Chassis Management Switch (CMS) provides connectivity to the BMC (Baseboard Management Controller) present on each server blade. There are eight 100Mbps Ethernet connections and one 1Gb Ethernet connection available in the CMS. Each slot has its own 100Mbps-dedicated Ethernet Interface connected to the blade BMC. The 1GbE interface is used to connect the CMS to the Redwood IO_MUX. The CMS is an unmanaged switch that requires no configuration.

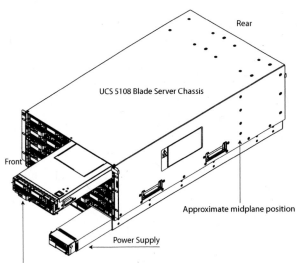

Figure 106: UCS 5108 Front Isometric view

5.2.3. *UCS 5108 Blade Server Chassis*

The UCS 5108 Blade Server Chassis is depicted in Figure 97, Figure 106 and Figure 107. The physical dimensions are reported in Section 7.2. On the front of the chassis are the eight half-slots for the server blades and the four slots for the power supplies. On the back of the chassis are the slots for the eight fans, the two Fabric Extenders and the power entry module. The airflow is front-to-back and the cooling is extremely efficient due to a chassis midplane that is

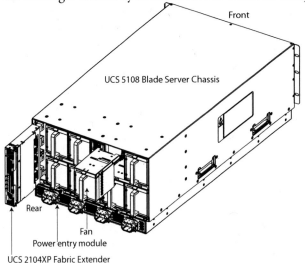

Figure 107: UCS 5108 Rear Isometric view

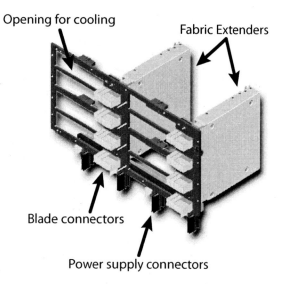

Figure 108: UCS 5108 Mid-plane

widely open and facilitates the airflow.

Figure 108 shows the chassis midplane that is installed vertically in the middle of the chassis, approximately 3/4 toward the back of the chassis (see Figure 106),

5.2.4. Common Blade Architecture

The UCS blades are based on Intel® reference architecture for dual-socket Nehalem-based servers (see Section 2.4.). These blades have multiple components in them and Figure 109 helps identifying each of them.

Figure 110 shows how the different components are interconnected.

The CPUs

The UCS blades use the Intel® Xeon® 5500 series CPUs. At the time of the first customer shipment four different CPU models can be installed on California. Table 8 illustrates the differences between them.

All CPUs have 4 cores and Hyper-Threading (see Section 2.1.4.) and TurboBoost (see Section 2.4.2.) capabilities. Intel® Nehalem architecture is described in detail in Section 2.4.

Intel® Xeon® L5520	
CPU speed	2.26GHz
Power Consumption	60W
QPI speed	5.87GT/s
Memory and maximum speed	DDR3@1066MHz
TurboBoost bin upside (133 MHz increments)	1/1/2/2

Intel® Xeon® E5520	
CPU speed	2.26GHz
Power Consumption	80W
QPI speed	5.87GT/s
Memory and maximum speed	DDR3@1066MHz
TurboBoost bin upside (133 MHz increments)	1/1/2/2

Intel® Xeon® E5540	
CPU speed	2.53GHz
Power Consumption	80W
QPI speed	5.87GT/s
Memory and maximum speed	DDR3@1066MHz
TurboBoost bin upside (133 MHz increments)	1/1/2/2

Intel® Xeon® X5570	
CPU speed	2.93GHz
Power Consumption	95W
QPI speed	6.4GT/s
Memory and maximum speed	DDR3@1333MHz
TurboBoost bin upside (133 MHz increments)	2/2/3/3

Table 8: Intel® Xeon® processors

The I/O Hub (IOH)

Both UCS blades use the Intel® X58 chip as an I/O Hub. This chip is connect-

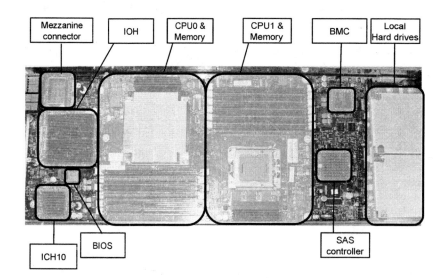

Figure 109: Blade components

ed to the CPUs with the Intel® QuickPath Interconnect (see Section 2.1.8.). The connectivity to the I/O Controller Hub is based on the Intel® Direct Media Interface (DMI). The Direct Media Interface is a point-to-point interconnection between the I/O Hub and the I/O Controller Hub. In previous Intel® architectures it was used as a link between the Northbridge and the Southbridge.

The X58 chip supports 36 PCIe lanes. In the UCS blades, these lanes are arranged in:

- One PCIe x4 link used for the SAS Controller;

- One PCIe x16 links used for the first mezzanine cards;

- One PCIe x16 links used for the second mezzanine cards (only on the UCS B250-M1).

The X58 PCIe ports support PCIe2.0 and are capable of running up to 0.5GB/s per lane. That gives maximum bandwidth for 8GB/s (64 Gbps) for each UCS blade mezzanine connector.

There are two QuickPath interfaces in a X58 chip and each of them is capable of running at speeds of 12.8GT/s. Each QuickPath Interface is connected to one CPU.

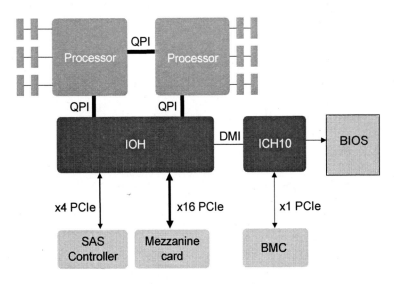

Figure 110: Blade block diagram

The I/O Controller Hub

UCS Blades use the Intel® ICH10 I/O Controller Hub. This chip is sometimes referred to as "Southbridge" (old name). The ICH is used to connect various "low-speed" peripherals, like USB devices, and it also provides connectivity for the BMC chip.

The BMC (Baseboard Management Controller)

In the UCS blades the BMC is used to provide pre-OS management access to blade servers and KVM (Keyboard Video and Mouse) access. It also functions as an aggregation point for the blade hardware.

The BMC used is a single chip, IP-based, server management solution. It provides the following functionality:

- General Inventory;
- Blade Thermal, Power and Health monitoring;
- KVM access to blade;
- Front panel video/USB access;
- Serial over LAN;
- Provide IPMI 2.0 interface to manage blade.

The BMC has two integrated 100Mb Ethernet connections that are connected in a redundant manner to the Chassis Management Switches inside the Fabric Extenders. The BMC connects to the ICH10 via a PCIe connection.

To provide the KVM function and the front panel video access, the BMC has an integrated graphics engine that is Matrox G200e compatible.

The SAS Controller

The UCS blades use the LSI Logic 1064e storage processor. It supports 1.5 and 3GB/s SAS and SATA transfer rates. It has integrated mirroring and striping functions to provide different RAID availability levels for internal disks in blades.

The Storage Controller is connected to the processing complex by x4 PCIe Gen1 connection to IOH module. Both UCS blades use Small Form Factor (SFF) drives. Normal SAS and SATA drives will be supported as well as SSD drives.

5.2.5.   UCS B200-M1 2 Socket Server

The UCS B200-M1 2 Socket Server (see Figure 111 and Figure 98) has all the features described in Section 5.2.4. as well as the following memory and mezzanine card configuration.

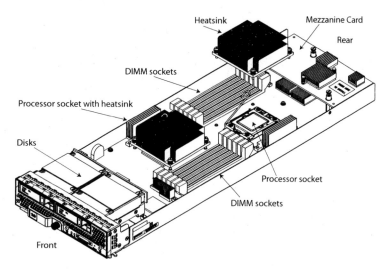

Figure 111: UCS B200-M1 Isometric view

Memory

The UCS B200-M1 blade supports up to 12 DDR3 DIMMs. Two DIMM slots are connected to each of the six memory channels. The supported memory speeds are 800, 1066 and 1333MHz. Not all speeds are available in all memory configurations. Table 9 lists the supported memory configurations (i.e., number of ranks and number of DIMMs per memory channel) and the resulting maximum memory speed.

Number of ranks	800 MHz	1066 MHz	1333 MHz
Single rank	2 DIMMs per channel	2 DIMMs per channel	1 DIMMs per channel
Dual rank	2 DIMMs per channel	2 DIMMs per channel	1 DIMMs per channel
Quad rank	2 DIMMs per channel	1 DIMMs per channel	Not supported

Table 9: Supported memory configurations

For example:

- twelve 8GB quad rank DIMMs can operate only at 800 MHz speed, even if they are rated at 1066 MHz;
- six of the same DIMMs (one DIMMs per memory channel) can operate at 1066 MHz.

To achieve the fastest possible speed (1333 MHz) only 6 DIMMs can be used (this reduces the load on the memory channel allowing higher speed). They have to be placed one per memory channel. If there is more than one DIMM in a memory channel the 1333MHz DIMMs are lowered to 1066MHz.

If different DIMM are mixed in a single blade, all the DIMMs will default to the same speed, i.e. the one of the lowest speed DIMM.

Mezzanine cards

The UCS B200-M1 blade has one mezzanine card slot available for one I/O Adapter. Available mezzanine cards are explained in detail in Chapter 4.

5.2.6. UCS B250-M1 Extended Memory Server

The UCS B250-M1 Extended Memory Server (see Figure 99 and Figure 112) has all the features described in Section 5.2.4. as well as the following memory and mezzanine card configuration.

Memory

The UCS B250-M1 blade has the Cisco® specific memory extension architecture described in Section 3.3. It is based on Cisco® ASICs that allows the UCS B250-M1 blade to address up to 4 times the memory of a standard Nehalem processor.

The UCS B250-M1 blade has 48 DIMM slots that can be fully populated and still operate at 1066MHz speed. From the BIOS, CPU or OS point of view these 48 DIMMs still look like 12 DIMMs and there is no need to modify or insert specific drivers in the Operating Systems or Applications. In fact this technology is totally transparent to the Operating Systems and Applications.

Mezzanine card

The UCS B250-M1 blade has two mezzanine card slots available for I/O adapters. Available mezzanine cards are explained in detail in Chapter 4.

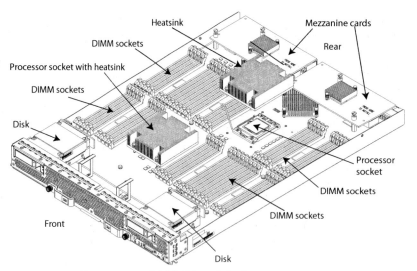

Figure 112: UCS B250-M1 Isometric view

5.3. Description of Communication Flows

The UCS system leverages consolidated I/O to communicate between different components in the system and it divides the traffic into three different categories: management, IP, and FC. The management traffic is out-of-band and leverages two pre allocated classes, dedicated for internal traffic. Separating the management traffic from other traffic ensures that management traffic is always available and communication is possible between the different components. Table 10 lists the different traffic types with reference to the connections highlighted on the UCS block diagram of Figure 113 (FEX is the Fabric Extender).

5.3.1. The Boot Sequences

The UCS systems components boot up when power is restored. During the boot phase the component is not responsive. Depending on the components the boot sequence goes through different steps.

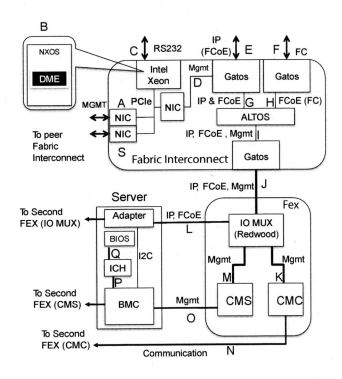

Figure 113: UCS block diagram

Description	Connection	Type
CLI (SSH, Telnet), GUI, KVM, XML API <-> DME	A, B	Mgmt
Console	C, B	Serial RS232
DME <-> Peer Fabric Interconnect	B, S	Cluster/Mgmt/Data
DME <-> IO_MUX	B, D, G, I, J	Mgmt
DME <-> CMC	B, D, G, I, J, K	Mgmt
DME <-> Adapter (Palo, Menlo)	B, D, G, I, J, L	Mgmt
DME <-> CMS	B, D, G, I, J, M	Mgmt
DME <-> BMC	B, D, G, I, J, M, O	Mgmt
DME <-> BIOS	B, D, G, I, J, M, O, P, Q	Mgmt
Adapter <-> Uplink Ether Port's (IP & FCoE)	L, J, I, G, E	Ethernet
Adapter <-> Uplink FC Port's (FC)	L, J, I, H, F	Fibre Channel
FEX <-> to peer FEX	N	Serial RS232

Table 10: Traffic types

POST

POST (Power-On Self-Test) runs when power is turned on.

It is a diagnostic testing sequence that a component (Processor) runs to determine if the hardware is working correctly.

Boot Loader

Boot Loader is a bootstrapping process.

The boot sequence is the initial set of operations that a computer performs when it is switched on, that loads and starts an operating system when the power is turned on.

Boot Errors

All components in the UCS that have two firmware banks will always try to boot from backup bank (Backup version) in case of a boot failure.

5.3.2. Fabric Interconnect and UCSM

The Fabric Interconnect has 2 different firmware images that need to be booted Kernel and System. The boot process is as follows:

1. POST (Power On Self Tests);

    o   Low level diagnostics;

2. Boot Loader loads;

    o   Boot Loader reads the boot pointer from flash memory and boots starting version (startup version defined by administrator) of Kernel;

3. Kernel loads and starts;

    o   Kernel components for NX-OS;

4. System loads and starts;

    o   System components (Port manager, etc);

    o   System plug-in UCSM;

    o   Self diagnostics;

5. Complete.

UCS Manager

The UCS manager (UCSM) is a plug-in to NX-OS. At boot time the NX-OS starts daemons and services, one of them is UCSM. The firmware repository is located on each Fabric Interconnect. If more versions of UCSM are available, only one can be started at a time: NX-OS will start the version the administrator chooses as the Startup Version.

5.3.3. Fabric Extender

The Fabric Extender (FEX) has two slots or banks with firmware, and a boot pointer. The boot pointer determines which of the two banks is the startup, the other one is declared the backup.

The boot process is as follows:

1.   POST (Power On Self Tests);

   o   Low level diagnostics;

2.   Boot Loader loads;

   o   Boot Loader reads the boot pointer from memory and boots the firmware from the startup bank/slot (Starting version);

3.   Firmware loads and starts;

   o   Internal system components start;

4.   Self diagnostics;

5.   Complete.

5.3.4. *Baseboard Management Controller*

The BMC has two slots or banks with firmware, and a boot pointer, just like the FEX.

The boot pointer determines which of the two banks is the startup, the other one is declared the backup.

The boot process is as follows:

1.   POST (Power On Self Tests);

   o   Low level diagnostics;

2.   Boot Loader loads;

   o   Boot Loader reads the boot pointer from memory and boots the firmware from the startup bank/slot (Starting version);

3.   Firmware loads and starts;

   o   Internal system components start;

4.   Self diagnostics;

5.   Complete.

6. UCS Manager

The UCS Manager (UCSM) provides a single point of management for the entire California system (aka UCS)[1]. It is an embedded policy-driven software that manages all devices in a California system as a single logical entity.

6.1. UCSM overall architecture

6.1.1. System Components

The architecture of UCSM consists of multiple layers with well-defined boundaries (see Figure 114). External interfaces provide communication with the outside world. The Data Management Engine (DME) is the central service that manages the components of a California system. Application Gateways act as a hardware abstraction layer between the DME and the managed end-points (EPs). The EPs are the actual devices or entities that are managed by UCSM but are not considered as part of UCSM itself.

External Interface Layer

The External Interface Layer consists of multiple interface components, for

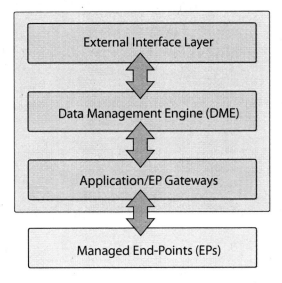

Figure 114: California UCSM components

---

1   The authors thank Mike Dvorkin for his contribution to this chapter.

outside world communication (e.g. SMASH-CLP, CIM-XML, and the UCSM-CLI), see Figure 115. The interface components are modular and well defined within UCSM. This design facilitates development of new interface components (to support new standards) or to enhance an existing interface component (to accommodate new features in currently-supported standards). All interface components are stateless in nature and their internal communication terminates on the DME.

Data Management Engine

The Data Management Engine (DME) is the central component in UCSM and consists of multiple internal services (see Figure 116). The DME is the only component in a California System that stores and maintains states for the managed devices and elements. The DME is the authoritative source of configuration information. It is responsible for propagating configuration changes to the devices and endpoints in the California system. The DME manages all devices and elements and represents their state in the form of managed object (MOs). MOs contain the desired configuration and the current state of a corresponding endpoint.

Administrators make changes to MOs; These changes are validated by the DME and propagated to the specific endpoint. For example, suppose an operator initiates a server "power on" request through the GUI. When the DME receives the request, it validates it, and if it is valid, the DME makes the corresponding state change on the server object in the Model Information Tree (MIT). This state change is then propagated to the server via the appropriate Application Gateway (AG).

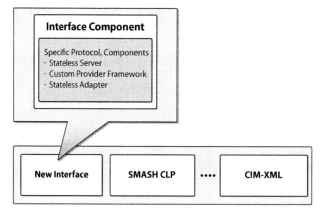

Figure 115: External Interface Layer

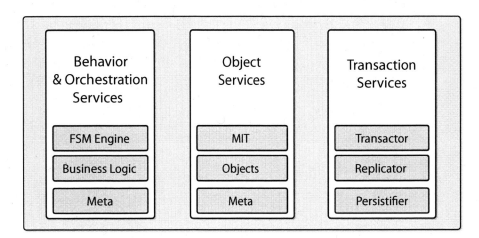

Figure 116: Data Management Engine Layer

Object Services

Hardware components, such as network interface cards, ports, servers, processors are represented as MOs. Statistics, faults, and events, are also represented as MOs. All MOs are hierarchically organized and contained in the MIT. The object services are driven by a meta-model that validates object properties, containments (objects that contain other objects) and relationships. They also handle the object life cycle.

Behavior & Orchestration Services

Behavior and orchestration services are driven by a meta-model and handle the behavior and rules regarding the objects. Each object contains meta-information that describes such things as the object properties, object type, parsing, rendering rules, inheritance and also how to invoke any business logic (product feature) specific to that object. Managed objects have their own Finite State Machine (FSM) in the form of a child object that is responsible for scheduling and performing tasks for the object. For example, a hardware inventory of a server is orchestrated by the FSM on the corresponding server object.

Transaction Services

Transaction services are responsible for the actual data mutations (configuration and state changes) of the objects (as performed by the "transactor" thread), and replication of state change to the secondary UCSM instance in a HA-environment (as performed by the "replicator" thread). It also verifies that the changes are permanently stored ("persistified") in the embedded persistent storage (as performed by the "persistifier" thread). Changes are made in an asynchronous and transactional fashion. No transaction artifacts are externally visible until data is persistified and replicated. This enables UCSM to provide great scalability and simultaneously guarantee a stable and consistent data model.

Application Gateways (AGs)

Application gateways are stateless agents that are used by the DME to propagate changes to the end-points. They also report system state from the end-points to the DME.AG is a module that converts management information (e.g.,configuration, statistics, and faults) from its native representation into the form of a managed object (MO). The AG is the hardware abstraction layer that abstracts the object model from the managed device or entity. AGs implement the platform-specific details of each of the managed endpoints. In UCSM, the AGs are implemented for the NX-OS, chassis, blades, ports, host agents, and NICs, see Figure 117.

- **Host AG.** The host AG is responsible for server inventory and configuration. It interacts with the server via the PNuOS (Processing Node Utility OS). This AG is responsible for BIOS firmware updates, local RAID controllers, third-party adapter option ROMs and also the RAID configuration of any local disks. It also ensures that the server becomes anonymous again by performing local disk scrubbing.

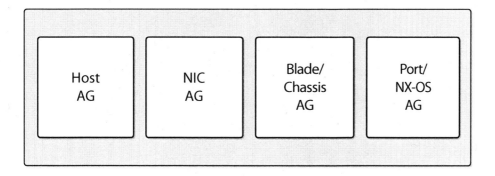

Figure 117: Application Gateway Layer

- **NIC AG.** The NIC AG monitors and manages the adapters, performs configurations changes, updates firmware and manages identifiers like MAC and WWN.

- **Blade/Chassis AG.** The Blade/Chassis AG monitors and manages the chassis management controller (CMC) inside the fabric extender (FEX) and the Baseboard Management Controller (BMC) located on the motherboard of each server. It performs firmware updates and configuration changes on the CMC and BMC, including UUID and BIOS settings. It also performs server-related operations like powering on and off the blades.

- **Port/NX-OS AG.** The Port/NX-OS AG monitors and manages the fabric interconnect. It performs actions such as configuring server-facing ports and Ethernet uplink ports. It is also responsible for VLANs port channels, trunks, and uplink connectivity for vNICs and vHBAs from the servers.

Managed Endpoints

Managed endpoints are resources within a California system that are managed by UCSM. The DME inside UCSM interacts with the endpoints via AGs. These resources are LAN-, SAN- and server- related, see Figure 118.

6.1.2. UCSM is a Model-driven Framework

UCSM uses a model-driven framework approach (see Figure 119) where system functionality is described in a generic, platform-independent information model (IM). "Content" is automatically generated by the NGEN (Nuova Generator, a program developed by Nuova Systems, now part of Cisco) from the Information Model (IM) and the manually coded "Business Logic", which extends and feeds into the Platform Definition Model (PDM). It is then automatically translated to a Platform Specific Model (PSM), described in C++. This generated code is then compiled and run inside the Fabric Interconnect and known as UCSM. About 70-75% of UCSM is automatically generated by code-generating robotics. Approximately 25-30% is written by developers and mainly consists of Business Logic and the specific product features.

When an operator initiates an administrative change to a California component (for example a boot request of a server) through one of the management interfaces, the DME first applies the change to the corresponding MO in the

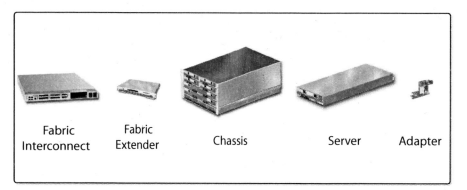

Figure 118: Managed End-Points

information model and then subsequently (and indirectly) the change is propagated and applied to the actual managed endpoint. The separation of business logic from platform implementation in a model-driven framework is useful for many reasons. One of them is the ability to develop the business logic independently from the platform implementation (and vice versa).

In UCSM, when the operator performs an action, the operator will get a response immediately, indicating that the DME has successfully received the change request. The immediate response is a success (if it is possible to satisfy the request) or a failure (if the request is impossible to satisfy). However, at

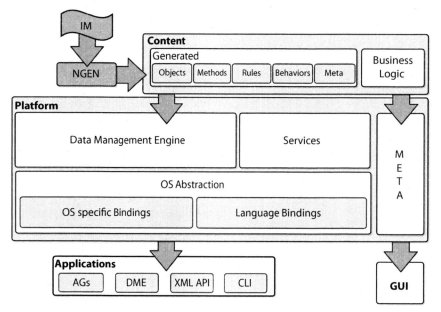

Figure 119: Model-Driven Framework

this time the End-Point may still not have yet received the request from the AG. For example, an operator can monitor the progress of a boot request by monitoring the "Power State". The "Power State" will change from "down" to "up" once the server is powered on.

Figure 120 and Table 11 contain an example of a server boot request.

Step	Command / Process	Administrative Power State of MO (Server)	Operational Power State of MO (Server)
1.0	CMD Request: Boot Server	Down	Down
2.0	Request gets Queued	Down	Down
3.0	State Change in Model Information Tree	Up	Down
4.0	Transaction Complete	Up	Down
5.0	Pass change info and boot request stimuli	Up	Down
6.0	Persistify the state change of MO to local store	Up	Down
6.1	Send state change information to peer DME	Up	Down
6.2	Persistify the state of MO to peer's local store	Up	Down
6.3	Reply with success (replication and persistification)	Up	Down
7.0	CMD: Response & External Notification	Up	Down
8.0	Apply reboot stimuli	Up	Down
9.0	Instruct BMC to power on server	Up	Down
10.0	Reply from BMC, server power on success	Up	Up
11.0	Reply, reboot stimuli success, Pass new Power state information	Up	Up

Table 11: Example of a Server Boot request

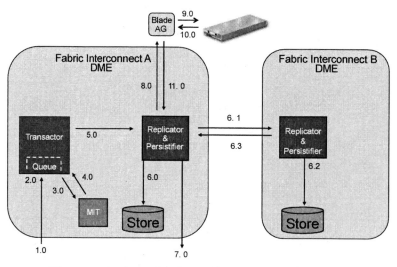

Figure 120: Sample Flow of Server Boot Request

UCSM lives in the Fabric Interconnect

UCSM is an NX-OS module and is therefore supervised, monitored, and controlled by the NX-OS, (see Figure 121). UCSM can be upgraded or restarted (just like any other module) without affecting I/O to and from the servers in the California system. However, the management will be unavailable during the time UCSM is down.

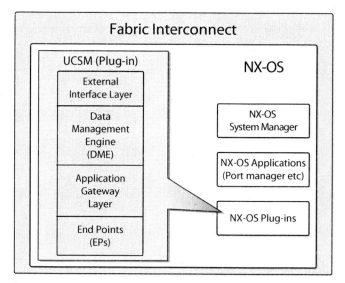

Figure 121: UCSM in the Fabric Interconnect

UCSM in a Highly Available configuration

UCSM can run in a highly available configuration. This is achieved by connecting two fabric interconnect devices together via two cluster ports on each fabric interconnect, see solid line in Figure 122. A fabric interconnect always runs an instance of UCSM, independently of the configuration. In a non-HA configuration, UCSM runs a standalone manager and controls all components of the California system. While the UCS management plane runs in an active-standby configuration, the data plane is active-active. Both fabric interconnects are actively sending and receiving network traffic, even though only one of them is running the active UCSM instance.

UCSM contains cluster components similar (but not identical) to a traditional active-passive application cluster. When two instances of UCSM are connected, they work together to achieve HA. There is an election after which one UCSM is promoted to "primary" and the other UCSM is demoted to "subordinate". The primary UCSM instance is the owner of a virtual IP address to which all-external management connections are made. The primary instance handles all requests from all interfaces and application gateways, performs model transformations (like configuration and state changes) in the system. The primary instance also replicates all changes in the system to the subordinate instance. This replication is done in an ordered and stateful way to eliminate the risk of partial information updates due to failure. This ensures that the subordinate instance always has a consistent and up-to-date state of all MOs in the system, in case of a failure of the primary instance. Both instances of

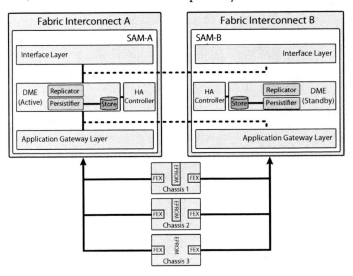

Figure 122: UCSM High Availability

UCSM are monitoring each other via the interconnect links. An odd number of chassis is used as quorum resources (writing on the chassis serial EEPROM) to determine the primary instance. For example in a California system with four chassis, three are used as quorum devices. The quorum devices also resolve a possible split-brain scenario which occurs if both cluster interconnect links are non-functional/disabled. Upon split-brain detection, both instances of UCSM will demote themselves to subordinate and at the same time try to claim ownership of all the quorum resources. The instance of UCSM that claimed most quorum resources will be the winner. This mechanism protects the California system from a situation where multiple UCSMs are running as active and believe they are the only active UCSM (split-brain).

6.2.   Management Information Model

As previously discussed, the UCS Management Information Model (MIM) is a tree structure where each node in the tree is a managed object (MO), see Figure 123. Managed objects are abstractions of real world resources – they

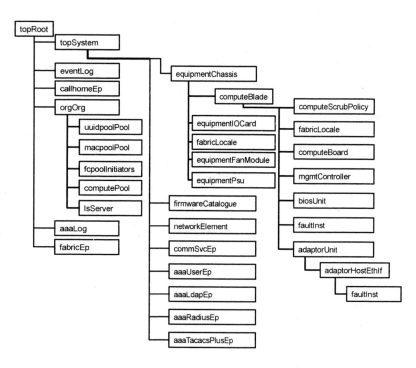

Figure 123: Managed Objects

represent the physical and logical components of the California system like fabric interconnect, chassis, servers, adapters, etc. Users do not create certain MOs. They are automatically created by the California system, such as power supplies and fan modules. The properties in the MOs are characterized as operational and administrative settings. At MO creation time, each MO is assigned a Dn (distinguished name) based on the value of its naming properties. Dn's lifecycles are coincident with the MO that they identify; That is, they are immutable once the MO is created. A Dn is a slash "/" delineated sequence of Rns (Relative names). For example, the Dn for a compute blade is "sys/chassis-1/blade-1". The Dn is used to unambiguously identify MOs and MO hierarchies that are the target of query and configuration operations.

Figure 123 shows containment relationships of MOs and how they are related in the MIM. Note that this is a small subset of the tree structure. Also note that fault MOs and statistic MOs are parented by the faulty and monitored MOs, respectively. Later this chapter describes other concepts like policies, service profiles, pools, templates that are also MOs in the MIM.

California allows users to query the state of MOs using Dns. It also supports hierarchical queries. A hierarchical query returns all the MOs in the MIT subtree rooted at the identified MO.

Figure 124 depicts an MO (in this example, a host Ethernet interface) in the MIM. The MO Dn is: "sys/chassis-3/blade-1/adaptor-1/host-eth-1."

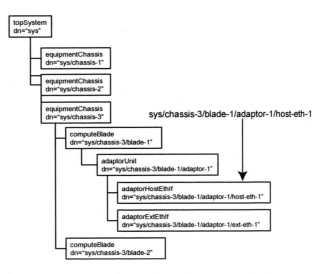

Figure 124: Information Model

6.3.   Available integration points

6.3.1.   *Interfaces*

All standard interfaces (see Figure 125) cover applicable subsets of the UCSM information model. The standard protocols that connect directly (cut-through) to a BMC (Baseboard Management Controller) via a unique external IP address bypasses UCSM. All other interfaces terminate on UCSM. Each request is queued, interpreted, de-duplicated, checked against the requestors' privileges, and executed by UCSM.

6.3.2.   *Standard (cut-through) interfaces in a UCS*

A cut-through interface provides access to a single server. The disadvantage of using such an interface is that it bypasses the DME. The DME is always in discovery mode, and it detects any changes made through a cut-through interface, like a reboot. The California system provides a unique external management IP address to each BMC for external management.

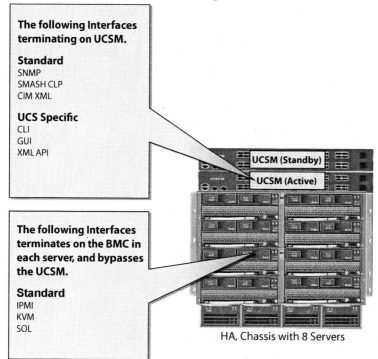

Figure 125: System Interfaces

IPMI (Intelligent Platform Management Interface)

IPMI is a protocol for monitoring and managing single server hardware information such as voltage CPU statistics and ambient temperature. Data collected from sensors are captured and available at the BMC. This interface is commonly used by management software to perform server management out-of-band, including reboot, power-on, and power-off. In a California system these capabilities are a small subset of all the capabilities provided natively by UCSM via the XML API.

SOL (Serial-over-LAN)

SOL enables an administrator to remotely connect to a single server and get full keyboard and text access to the server.

KVM (Keyboard-Video-Mouse)

KVM enables an administrator to remotely obtain keyboard, video, and mouse access to a single server, it is commonly used to install a single server OS and to

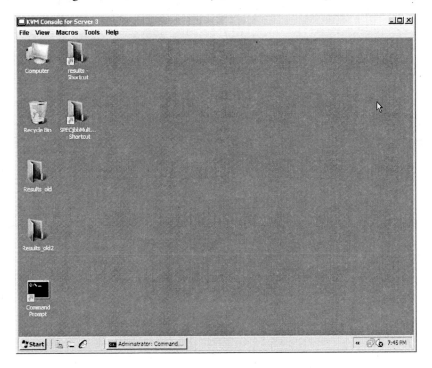

Figure 126: KVM session

troubleshoot OS-related issues (see Figure 126).

KVM also enables for virtual media, which allows an administrator to mount a remote media on the server the KVM is accessing. The administrator can give the server access to image files (ISO) or physical drives like CD and DVD players on the client running the KVM console session.

The KVM console can be launched from the UCS manager GUI or as a stand-alone application. The session and the user is always authenticated by the UCS manager independently from where the KVM session was launched. The stand-a-lone KVM is downloadable from the UCS manger via a URL. The stand-a-lone KVM application connects to external management IP addresses assigned to each BMC allocated from an IP address pool (External IP management Pool).

A KVM console launched from within the UCSM GUI it does not require an external management IP address.

6.3.3.   Standard Interfaces in a UCS

SNMP (Simple Network Management Protocol)

The California system provides a read-only SNMP interface for monitoring of the fabric interconnects. This interface in the UCS in not designed to be used for monitoring the chassis, servers, or adapters.

SMASH-CLP (Systems Management Architecture for Server Hardware Command Line Protocol)

The SMASH-CLP enables administrators to use a standard command-line interface for servers, independent of vendor and model. In the UCSM implementation, this interface provides read-only access and can be used for monitoring and debugging, inventory collection, and other such information gathering of servers and chassis.

CIM-XML (Common Information Model-eXtensible Markup Language)

The CIM-XML standard interface defines a Common Information Model for servers, which allows software to programmatically exchange this well-defined information with different vendor systems. In a California system, this interface is implemented as read-only and can be used for monitoring, debugging and inventory collection (server and Chassis related) by frameworks that support the CIM-XML standard.

6.3.4. Native Interfaces in California

UCSM CLI (UCSM Command-Line Interface)

The command-line interface (CLI) is one of several interfaces you can use to access, configure and monitor the California system. The CLI is accessible from the console port or through a Telnet or SSH session. The UCS CLI is object based and fully transactional.

UCSM GUI (UCSM Graphical User Interface)

The UCSM GUI (see see Figure 127) is JAVA based, it can be started and run from any supported web browser. This interface provides a fully featured user interface for managing and monitoring of a California system.

XML API, eXtensible Markup Language-Application Program Interface

The XML API is the most powerful interface used to integrate or interact with a California system. It is generic, content driven and hierarchical. This is the native language for the DME and therefore there are no restrictions on what can be done through this interface, within the feature of the DME. The XML API also supports event subscriptions which allow a subscribing client, like monitoring software, to receive all events or actual state changes in the whole California system, assuming that the subscriber has the right privileges. This is useful as the feed only sends actual changes and not whole MO information; there is no need for the client to poll for current state and then differentiate what may have changed since last pull for information.

All interfaces including UCSM GUI and UCSM CLI interfaces (except cut-through interfaces) are translated into the native XML API prior to reaching the DME, see Figure 128. Please note that the SNMP interface is not included in the picture, even though the SNMP interface is configured by UCSM, it is a native interface in the fabric interconnects.

Sample Integrations

Figure 129 is a sample of how the interfaces in the UCS can be leveraged.

All the interfaces benefit from the fact that the DME is transactional, since all

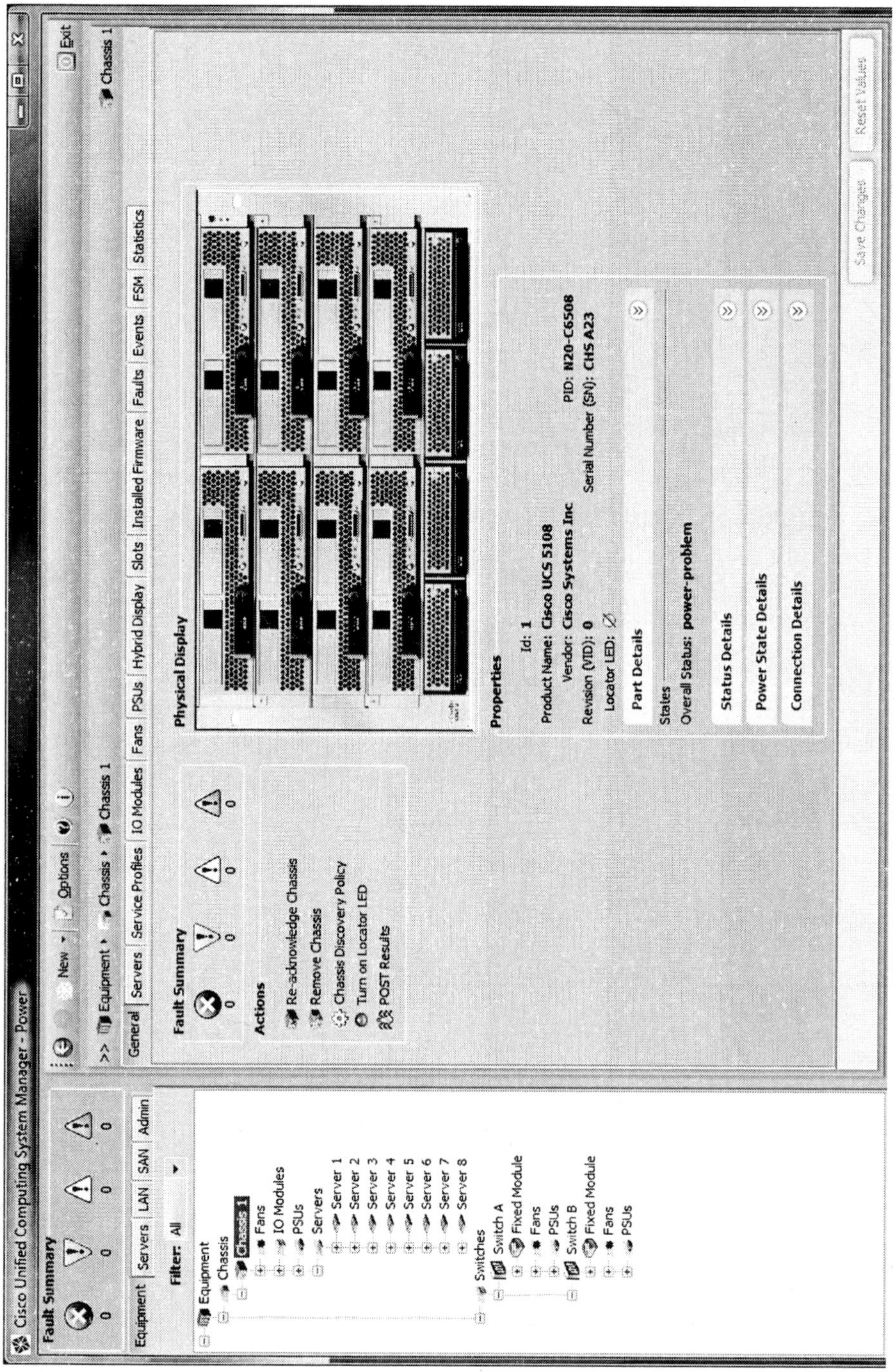

Figure 127: GUI

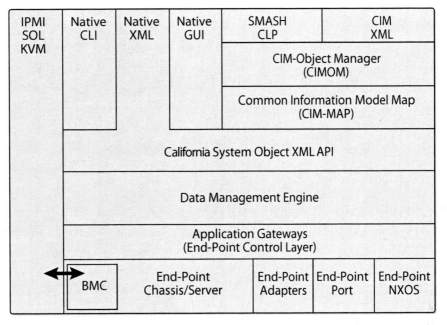

Figure 128: System Interface Stack

the interfaces are translated into native XML API and the requests handled by the DME. This also guarantee uniform enforcement of Role Based Access Control (RBAC), etc.

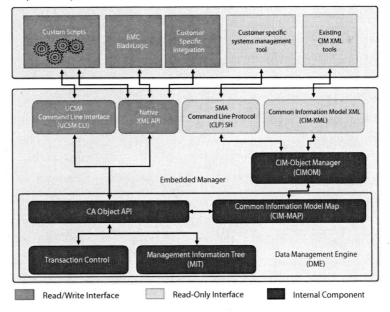

Figure 129: Example of Integrations and Interfaces

Syslog

Syslog is a standard protocol for forwarding log messages in a client/server fashion. The log messages include information from all the components in the California system. The California system allows the ability to forward log messages to up to three different syslog hosts or to a local Console, Monitor or File (see Figure 130).

Operating principles

UCSM is a policy driven management device manager. Policy-driven management is one of the key features of UCSM, and it helps IT organizations to better define and implement their own best practices.

Policies help to ensure that consistent, tested and compliant systems are used, reducing the risk of issues caused by repetitive manual tasks. Data centers are becoming increasingly more dynamic, and the definition, consumption, and resolution of policies is a key enabling technology for making infrastructure devices portable and reconfigured dynamically. The California system allows subject matter experts like network, storage and server administrators to pre-define policies within their area of expertise, which can later be selected by a server administrator when defining a compute resource.

Policies control behaviors and configurations. They control how a server or other components of the Cisco California system will act or be affected in specific circumstances. The California system has a large number of different policies. These policies describe and control all aspects of the California sys-

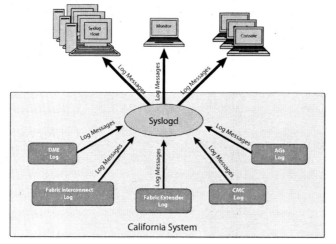

Figure 130: Syslog Messages Flow

tem such as network, storage, server configurations and system behavior.

The California system is managed by rules and policies. Rules are defined in form of policies inside the UCSM, and enforced in the end-point (devices). This removes states from the devices and improves mobility and scalability. Policies are centrally defined and enforced at the end-point, they can be used to perform policy enforcement on any device within the same California system. As everything in a California system, policies are managed objects and as any other MO can be exported and imported to other California systems.

6.3.5. Configuration Policies

Configuration policies are the majority of the policies in a California system and are used to describe configurations of different components of the system. Table 12 contains some policy examples valid at the time of this writing.

Policy	Description
Autoconfiguration Policy	describes how to automatically configure a newly discovered server.
Boot Policy	determines the location from which a server boots, like via a SAN, LAN, Local Disk, or Virtual Media.
Chassis Discovery Policy	determines how the system reacts when a new chassis is discovered.
Dynamic Connection Policy	determines how the VNTag connectivity between VMs and dynamic vNICs are configured.
Ethernet Adapter policy	determines how an Ethernet adapter handles traffic (such as queue depth, failover timeout, performance, etc).
Fibre Channel Adapter policy	determines how a Fibre Channel adapter handles traffic (includes flogi & plogi timeout, error handling, etc).
Host Firmware Pack Policy	determines a common set of firmware versions that will be applied to a server via a service profile.
Local Disk Configuration Policy	determines RAID and configuration on the optional local drives.
Management Firmware Policy	determines a firmware version that will be applied to the BMC on the server.

QoS Definitions Policy	defines the outgoing quality of service (QoS) for a vNIC or vHBA (class of service, burst and rate).
Server Discovery Policy	determines how the system reacts when a new server is discovered.
Server Inheritance Policy	describes the inheritance from the hardware on a newly discovered server.
Server Pool Policy	determines pool memberships of servers that match a specific "server pool policy qualifications".
Server Pool Policy Qualifications	qualifies servers based on inventory rules, like amount of memory, number of processors .
vHBA and vNIC Policy	defines connectivity and quality of a vHBA or vNIC.

Table 12: Configuration policies

6.3.6. Operational Policies

Operational policies determine how the system behaves under specific circumstances. Table 13 contains some policy examples valid at the time of this writing. Figure 131 contains an example of server profile policies.

Policy	Description
Adaptor Collection Policy	defines intervals for collection and reporting statistics regarding adapters.
Blade Collection Policy	defines intervals for collection and reporting statistics regarding blades.
Call Home Policies	defines how call home profiles are used for email notification.
Chassis Collection Policy	defines intervals for collection and reporting statistics regarding chassis.
IPMI Profile	defines the IPMI capabilities of a server and whether the access is read-only or read-write.
Fault Collection Policy	defines clear action and intervals for fault clearance and retention.
Port Collection Policy	defines intervals for collection and reporting statistics regarding ports.

Scrub Policy	determines if any state of the server should be kept during the discovery process.
Serial over LAN Policy	defines the serial over LAN capabilities of a server.
Threshold Policy	sets alarm triggers for Ethernet, Fibre Channel, adapters, blades, chassis, PSUs, FEXs, FANs, etc.

Table 13: Operational policies

6.3.7. Global vs. Local Policies

Some policies are global and some are local. The global policies are defined for the whole California system, local policies are on per organization level (organizations will be covered later in this chapter). Examples of global policies include Fault Collection Policy, Call Home Policies, and the various statistic Collector Policies.

Default vs. User created Policies

The California system has default policies that are used in the absence of other user specific created policy, the pre-defined default policies are typically defaulting to normal system operation behavior.

6.3.8. Pools

Pools are containers for resources and ID definitions, see Figure 132. There are mainly two different types of pools, Identity pools and blade pools. A pool contains only one type of resource, for instance, a blade pool can only contain blades and a WWN Pool can only contain WWNs. Pools make a system easier to manage and maintain since it minimizes the need for administrators to actively manage the use of resources. Multiple pools with different names can contain the same type of resources, but also the very same resources. For example, a blade can be a member of multiple pools. Once a blade is consumed and assigned by the system, the blade is no longer available to any pools. Pools are providers to Service Profiles and Service Profile Templates (discussed later in this chapter).

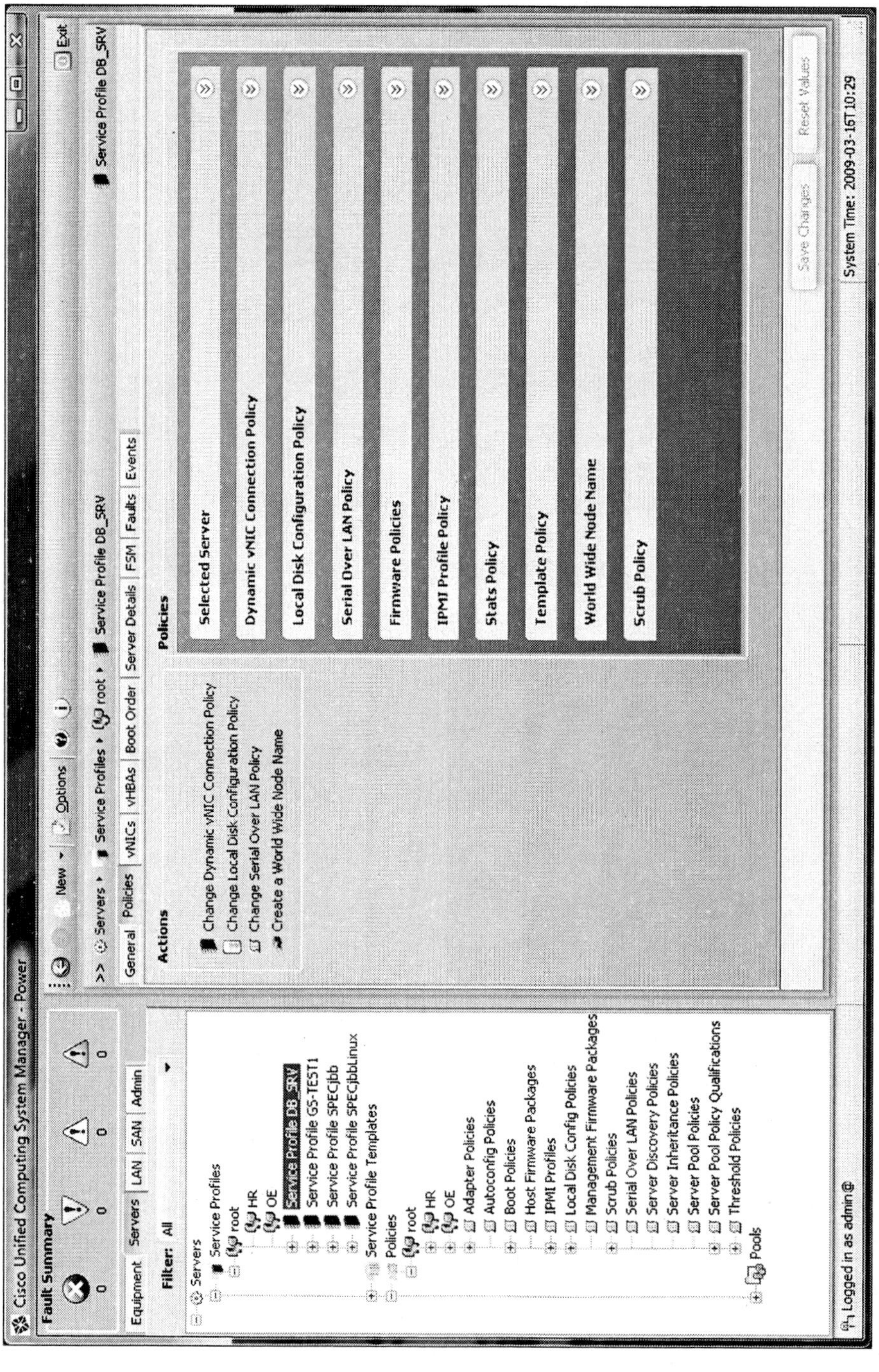

Figure 131: Service Profile Policies

Identity Pools

Identity pools are used for server identity management and assignments. This allows the California system to automatically assign identities to a server and their interfaces in the California system. This minimizes the administrative tasks needed for network, storage and server administrators. More importantly, this is one of the key components for stateless computing since the servers are stateless and any server can be assigned to any ID and workload at any time. There are three types of identity pools: MAC, WWN and UUID. Figure 132 shows an example of system pool types.

MAC Pools

A MAC (Media Access Control) address is a hardware address that uniquely identifies each node of a LAN. A MAC pool contains blocks of MAC addresses and gets populated by an administrator who creates one or more blocks of MAC addresses in the pool. These are sequential and the block assignment is done with a starting MAC address "From" and a number of addresses the block should contain (size of block). For example, a block could be as follows: From: 00:01:42:00:00:01 Size: 5. The block would look like Table 14. Figure 133 shows a screen capture of the MAC pool page.

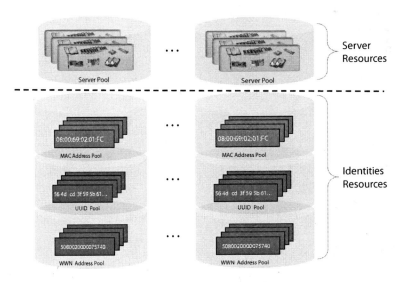

Figure 132: System Pool Types

00:01:42:00:00:01
00:01:42:00:00:02
00:01:42:00:00:03
00:01:42:00:00:04
00:01:42:00:00:05

Table 14: Example of MAC address pool

UUID Suffix Pool,

A Universally Unique Identifier (UUID) is an identifier that uniquely identifies each compute resource (server). A UUID pool contains blocks of suffix UUIDs. In a California system, a suffix UUID is 8 bytes and consists of 16 hexadecimal characters and has the format of "HHHH-HHHHHHHH". The pool is prefixed with a UCS unique 8 bytes in the format of "FFFFFFFF-FFFF-FFFF". The UUID assigned to the server is a "Prefix"-"Suffix", like "FFFFFFFF-FFFF-FFFF- HHHH-HHHHHHHH". The suffix UUIDs are populated by an administrator who creates one or more blocks of UUID numbers in the pool. From: "1000-0A0B0C0D01" To: "1000-0A0B0C0D05". Assuming the prefix is "1FFFFFFF-2FFF-3FFF", the block would look like Table 15.

1FFFFFFF-2FFF-3FFF-1000-0A0B0C0D01
1FFFFFFF-2FFF-3FFF-1000-0A0B0C0D02
1FFFFFFF-2FFF-3FFF-1000-0A0B0C0D03
1FFFFFFF-2FFF-3FFF-1000-0A0B0C0D04
1FFFFFFF-2FFF-3FFF-1000-0A0B0C0D05

Table 15: UUIDs

WWN Pools

A World Wide Name (WWN) is an address that uniquely identifies either a node or a port in a SAN. A WWN pool contains blocks of WWN addresses, very similar to the MAC pools. An administrator, who creates one or more blocks of WWN addresses in the pool, populates it. These are sequential and the block assignment is done with a starting WWN address "From" and a number of addresses the block should contain (size of block). For example a

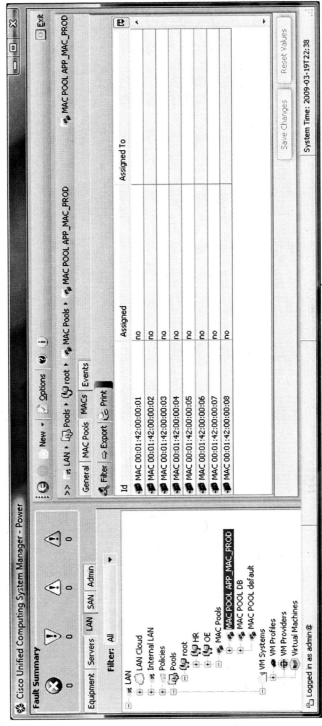

Figure 133: Mac Pool

block could be created as follows: From: 20:00:00:01:42:00:00:01 Size: 5 The block would look like Table 16.

20:00:00:01:42:00:00:01
20:00:00:01:42:00:00:02
20:00:00:01:42:00:00:03
20:00:00:01:42:00:00:04
20:00:00:01:42:00:00:05

Table 16: WWNs

A WWN in the pool can also be assigned a boot target and boot LUN, as in Table 17.

NOTE: All servers in this example boot from the same WWN and LUN "0". This is because storage arrays mask the real LUN IDs by using a translation layer between the back-end LUN ID and the server presented LUN ID. So the result is that all boot LUNs are physically different in the storage array but presented to the different servers as LUN ID "0".

WWN	Boot Target WWN	LUN ID
20:00:00:01:42:00:00:01	50:00:00:01:42:60:00:A1	0
20:00:00:01:42:00:00:02	50:00:00:01:42:60:00:A1	0
20:00:00:01:42:00:00:03	50:00:00:01:42:60:00:A1	0
20:00:00:01:42:00:00:04	50:00:00:01:42:60:00:A1	0
20:00:00:01:42:00:00:05	50:00:00:01:42:60:00:A1	0

Table 17: Boot target

Server Pools

A server pool contains a set of available, stateless servers. This pool type allows for two different ways to populate the members of the pool: manual and automatic.

6.3.9. Manual Population of pools

Manual population is a very simple but not very efficient way to assign pool membership for a server, see Figure 134. The operator manually maintains and manages the server membership. In Figure 134 an administrator is manually evaluating the inventory of each server and then assigns the server to one or more server pools. The picture shows two different pools (Application_A and Application_B) with the same requirements of server characteristics. So the administrator manually makes them members of both pools.

6.3.10. Automatic population of pools

A Pool can be automatically populated via policies, see Figure 135. There are two different policies that together decide which pool a server should be a member of: "Server Pool Policy Qualifications" and "Server Pool Policy".

A "Server Pool Qualifications" policy describes the hardware qualification criteria for servers, like number of Processors, amount of RAM, type of adapters, etc. A server that fulfills all of the defined criteria in the policy is considered a qualified server.

A "Server Pool Policy" describes which "Server Pool" the server becomes a member of, if it has qualified for a certain "Server Pool Qualification". The same server can meet multiple qualifications. Multiple "Pool Policies" can point to the same pools. Figure 136 shows the creation of a server pool policy

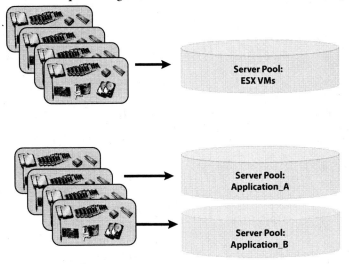

Figure 134: Manual Pool Assignment

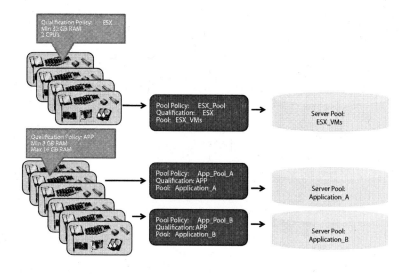

Figure 135: Automatic Pools Assignment

qualification and Figure 137 shows the creation of a server pool policy.

Figure 135 shows one qualification (APP) that has qualified multiple servers and two "Pool Policies" that are using the very same qualification to populate different "Server Pools".

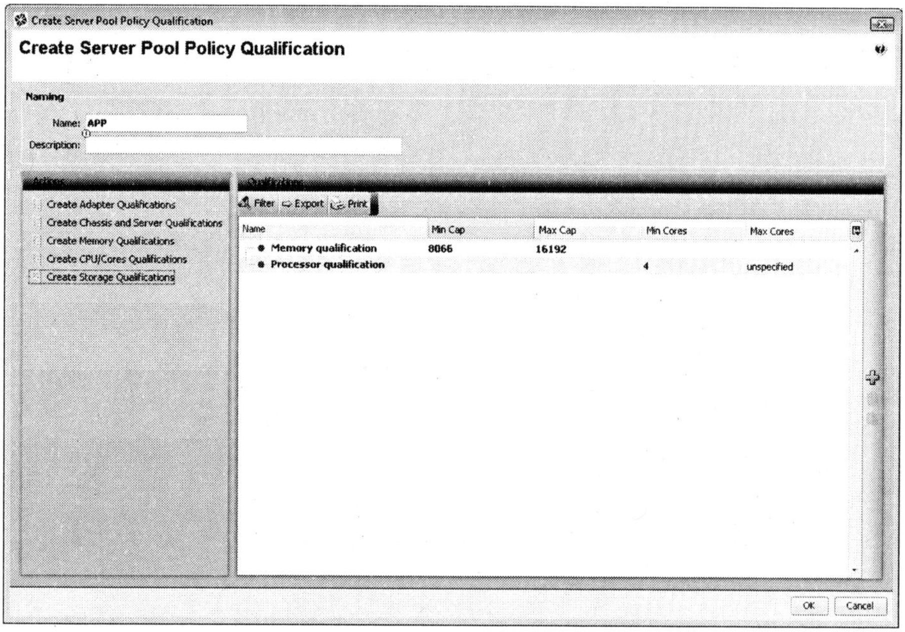

Figure 136: Server Pool Qualification

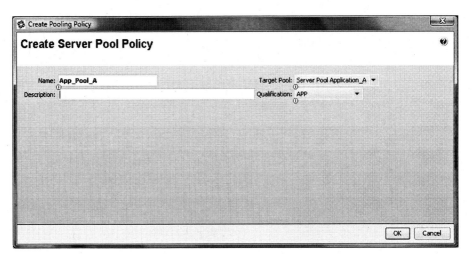

Figure 137: Server Pool Policies

Default vs. User created Pools

The California system also has default pools that are pre-defined, but not pre-populated. The default pools are used as a last resource when a regular pool has been drained out in the local organization, and there are no other resources available in parent or grandparent's organizations.

6.3.11. Service Profiles

A service profile is a self-contained logical representation (object) of a desired physical server, including connectivity, configuration and identity. The service profile defines server hardware (configuration, firmware, identity and boot information), fabric connectivity, policies, external management, and high-availability information, see Figure 138.

Every server is stateless and must be associated with a service profile to gain its identities and personality. A service profile is associated with a stateless server via manual association, or automatically via a blade pool.

A service profile ensures that the associated server hardware has the configuration, identities and connectivity to a LAN and SAN based on the requirements from the applications the server will host.

An associated server (with a service profile) is similar to a traditional bare metal server in the way that it is ready to be used for production and run business services or software. Once the service profile is de-associated from the server,

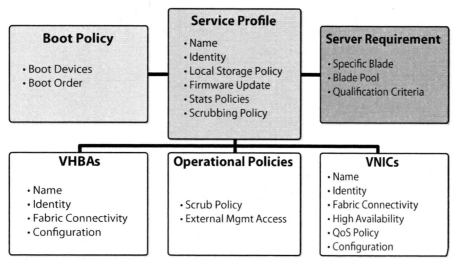

Figure 138: Service Profile Components

the server is reinitialized (erased from any state) and returned to the pool (if originated from a pool).

Servers and service profiles have a one-to-one relationship. Each server can be associated with only one service profile (see Figure 139). Each service profile can be associated with only one server. A service profile can be modified, cloned or used to instantiate a template.

The California system uses pools, policies and service profiles to abstract states and configuration information from all the components that define a server, service quality and its connectivity. This is one of the key areas that make the California system a stateless computing system. It allows separation of an operational state and services away from the server hardware and physical connectivity.

For example a service profile can be associated to any available server in a California system which automatically includes full migration of Identities, firmware, and connectivity to LAN and SAN, etc. Figure 140 summarizes the information that is included in a service profile.

6.3.12. Service Profile Templates

A service profile template is very similar to a service profile except it cannot be associated with a physical server. It can, however, be used for instantiation of multiple service profiles. This is useful for operators that need to instantiate

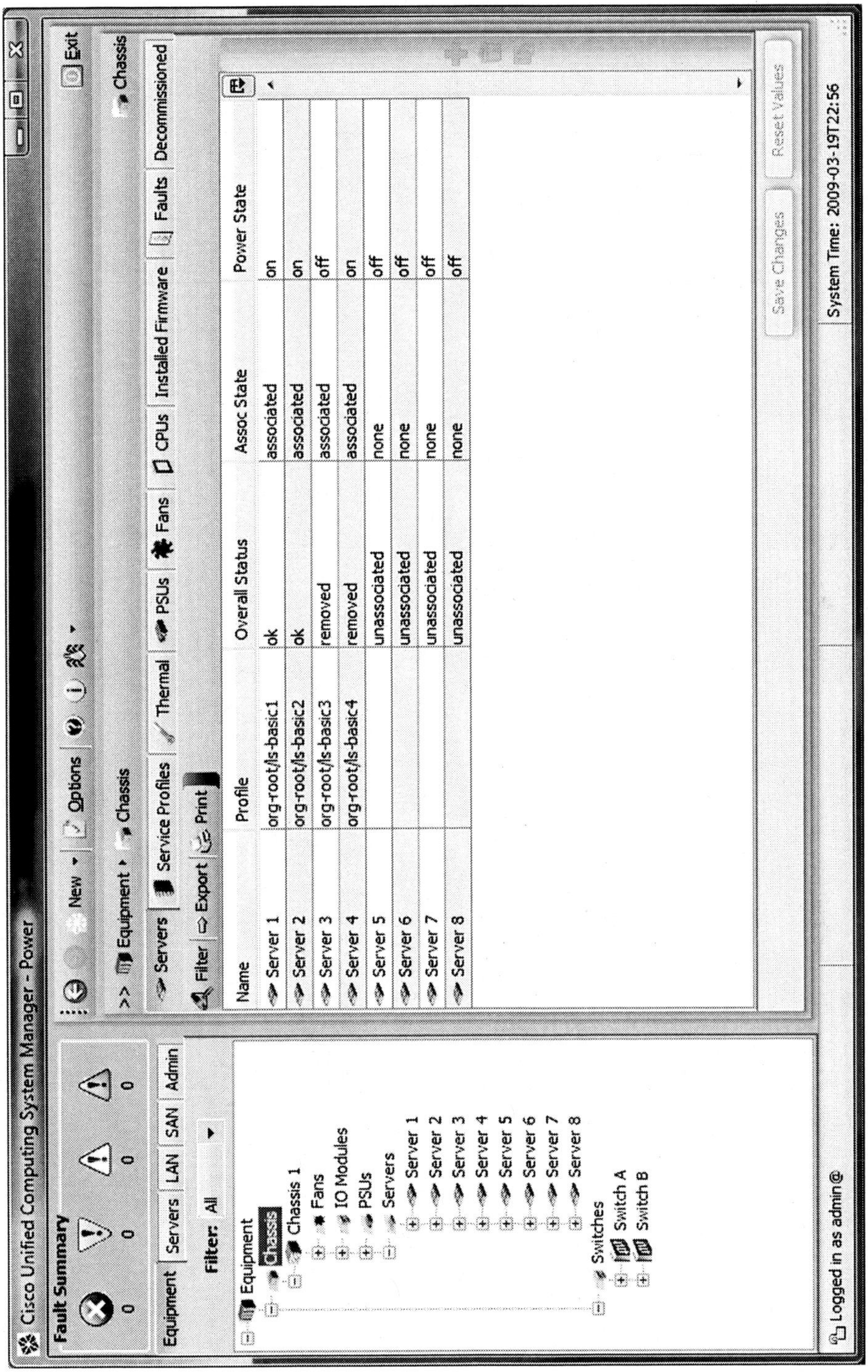

Figure 139: Service Profile Association

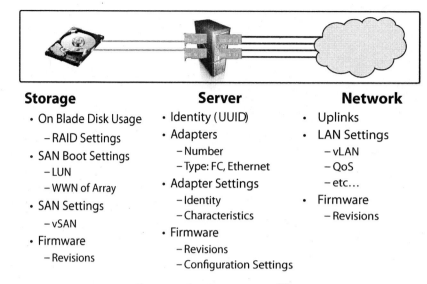

Storage
- On Blade Disk Usage
  - RAID Settings
- SAN Boot Settings
  - LUN
  - WWN of Array
- SAN Settings
  - vSAN
- Firmware
  - Revisions

Server
- Identity (UUID)
- Adapters
  - Number
  - Type: FC, Ethernet
- Adapter Settings
  - Identity
  - Characteristics
- Firmware
  - Revisions
  - Configuration Settings

Network
- Uplinks
- LAN Settings
  - vLAN
  - QoS
  - etc...
- Firmware
  - Revisions

Figure 140: Service Profile

a large number of service profiles. A senior server administrator can create a template for a certain purpose, like a Database template, or Application Template. This enables other administrators to instantiate new service profiles with little or no expertise of requirements for a certain application or software.

There are two types of Templates: "initial-template" and "updating-template". The difference is that a service profile created from an "initial-template" does not inherit any new configuration changes made to the template after the service profile has been instantiated. A service profile that has been created from an "updating-template" maintains relationship and keeps inheriting any configuration changes made to the template.

6.3.13. Organizations

In a California system, multi-tenancy can be accomplished by dividing up the large physical infrastructure of the system into logical entities known as organizations. As a result, administrators can achieve a logical isolation between organizations without providing a dedicated physical infrastructure for each organization. The structure of the organizations in a multi-tenancy implementation will depend upon the business needs of the company. For example, organizations that represent a division within a company, such as marketing, finance, engineering, human resources, or other organizations that represent a different customer, etc.

The "base" organization is root and all other organizations are hierarchical. All policies and other resources that reside in a root organization are system-wide and available to all organizations in the system. However, any policies and resources created in other organizations are only available to organizations in the same hierarchy. For instance, consider a system with organizations named "Finance" and "HR" that are not in the same hierarchy. "Finance" cannot use any policies from the "HR organization", and "HR" cannot use any policies from the "Finance" organization.

Unique resources can be assigned to each tenant, or organization. These resources can include different policies, pools, and quality of service definitions, etc. Administrators can also be assigned strict privileges by an organization.

6.3.14. Hierarchical Pool and Policy resolution

The way that UCSM resolves a pool or policy name to a service profile is from the bottom to the top of the organization tree. Consider the organizational structure in Figure 141. If a user creates a service profile in Org-D and associates a boot policy called policy-1 with it, since UCSM resolves policy names by traversing the tree upwards, the policy association for the service profile will be the policy-1 policy in the root organization. However, if there was a boot policy named policy-1 in Org-D, then UCSM will use that one and not the one in the root organization. Note that, in the absence of any user-defined boot policies, then the default boot policy associated with the organization will be used.

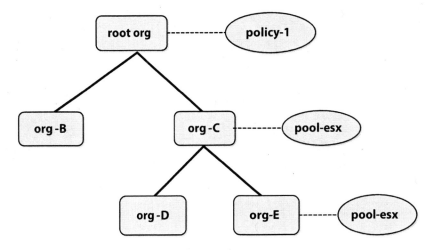

Figure 141: Hierarchical Pool and Policy resolution

Another nuance in the resolution of a pool to a service profile is that UCSM will search the tree for an available resource. For example, suppose a server pool named pool-esx is associated with a service profile in the Org-E. When that service profile is deployed, UCSM will search for an available blade from a blade pool called pool-esx from the org-E organization in the scope of parents and grandparents. If there is no available blade in the org-E pool but one available in the org-C pool, then UCSM will use the blade from the org-C blade pool. Just as with policies, in the absence of a user-defined server pool, the default server pool associated with the organization will be used.

In general, resources and policies are not strictly owned by any particular organization but instead are shared across all organizations in a sub-tree using a rule of "first-come first serve".

6.3.15. *Role-Based Access Control (RBAC)*

Role-Based Access Control (RBAC) is a method of restricting or authorizing system access for users based on "roles" and "locales". A role can contain one or more system privileges where each privilege defines an administrative right to a certain object or type of object (components) in the system. By assigning a user a role, the user inherits the capabilities of the privileges defined in that role. Customers can create custom specific roles (see Table 18).

6.3.16. *Locales*

A locale (not to be confused with the internationalization of character sets) in UCSM is designed to reflect the location of a user in an organization. A user is assigned all the privileges of his/her roles to all objects in the locale for the entire organizational sub-tree where the locale is associated. A locale describes where the privileges from the role can be exercised. By assigning an administrator a certain user role and a locale, the administrator can exercise his privileges in the organizations and sub-organizations defined in the locale. Because users are not directly assigned privileges, management of individual user privileges is simply a matter of assigning the appropriate roles and locales. For example, Figure 142 shows server related R/W privileges in each org. Based on Role and Locale, user "Bob" gets server administrative rights (R/W) in org "HR" and Read-Only (R) in "OE" and "Root" for server related tasks, all non server related tasks are Read-Only (R) in all organizations.

Role Name	Responsibilities	Privileges
server	Provision blades	Create, modify, and delete service profiles
network	Configure internal and border LAN connectivity	Create, modify, and delete port channels, port groups, server pinning, and VLANs
storage	Configure internal and border SAN connectivity	Create, modify, and delete port channels, port groups, server pinning, and VSANs
AAA	Configure authentication, authorization, and accounting	Create, modify, and delete users, roles, and locales, with the exception of "admin" role.
admin	Any and all	Any and all. Only a user with "admin role", can create other users with "admin role".
read-only	None	Read-only access to all status and configuration

Table 18: Example of roles

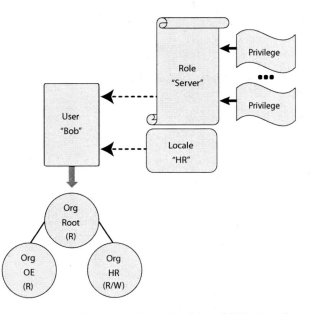

Figure 142: Server Admin Role and HR Locale

6.3.17. Users and Authentication

UCSM supports the creation of local users in the UCSM database as well as
the integration of name services such as LDAP, RADIUS, and TACACS+ for
remote users. When a user logs into the California system manager UCSM,
they are authenticated against the appropriate back-end name service and assigned privileges based on its roles.

A user assigned the "server role" performs server related operations within the
California system. A user assigned the network role has network privileges and
managed network related tasks. The storage role performs SAN-related operations. The scope of the AAA role is global across the California system. The admin role is the equivalent of the root user in a UNIX environment. The admin
role has no restrictions in terms of which privileges it has on resources.

6.4. Basic system management with UCSM

In a California system, UCSM is the embedded device manager and is the only
component that contains states. All other hardware components in a California system are stateless (except for local disk information that might contain
OS and application-related states). A California system discovery is an ongo-
ing process. UCSM is always looking for new components and changes to the
existing hardware configuration like insertion and removal of server, fans, power supplies, cables, etc. During the initial setup and during normal operation
of the California system, an administrator can configure and change desired
configurations, for example, HA vs. Non-HA configuration and what ports
on the Fabric interconnect are in use for servers and infrastructure. The California system knows the desired configuration, for example, what ports can be
connected to a chassis and what ports can be connected to the infrastructure.
UCSM will leverage this information to accept or reject illegal configurations.
Based on a desired configuration, UCSM will discover all components in the
system automatically.

6.4.1. H/W Management

There are two ways UCSM can be updated on hardware configuration changes, by a discovery process typically triggered by a sensor in the system such as
a presence sensor or by a link up state change on a port. It is also possible that
information changes when an AG updates the UCSM. UCSM can ask for

information from the AG, or AGs can be instructed by the UCSM to push information to UCSM when there are state changes on end-points.

6.4.2. Example of a Chassis Discovery Process

When a new chassis is automatically discovered the slots, power supplies, fabric extenders, fans, etc. are all inventoried and the chassis is assigned a chassis ID. Once the chassis is discovered and accepted, UCSM will continue to discover the presence of servers in each slot. If the presence of a server is discovered in a slot, UCSM will discover the model number, serial number, memory, number of processors, etc. UCSM will also perform a deep discovery of the server via PNuOS process (Pre-boot execution environment OS, it runs on a server to perform diagnostics, report inventory, or configure the firmware state of the server) that also functions as a validation of components.

Chassis Discovery

- Link on a server facing port is discovered;
- Open basic communication to CMC (Chassis Manager Controller) in the FEX;
- Check component compatibility;
- Check that this device is supported;
- Checks the Serial Number on Chassis, if it does exist:
    * Check if this device has been removed from management by the administrator?
    * This is a new link to existing chassis;
- If a new chassis it will accept it;
- Perform discovery, such as Model, firmware, Fans, PSUs, slots, etc.

Discovery of Server in Chassis

- If a slot raises a state then the slot has presence (a server is present);
- Opens communication with BMC on the server in that slot;
- Discovers Server information, such as:
    * Vendor, Model, SN ,BIOS, BMC, CPUs, Memory, HBAs,

NICs, and Local Storage Controller;

- Boot into PNuOS

- Deep discovery via PNuOS, such as:

  * Local disk info, Vendor specific HBA & NIC information;

- Enforce scrubbing policies of server.

6.4.3. Retirement of hardware

Hardware can be retired by using the "Remove" function in the system. Removed components are removed from UCSM MIT (Internal Database) and, therefore, are not present in the MIM. However, UCSM does maintain a list of decommissioned serial numbers just in case an operator mistakenly reinserts a removed component. An administrator can bring decommissioned components back into control under the UCSM.

6.4.4. Firmware Management

Firmware management - the operations that consist in keeping track of versions, dependencies, performing updates, maintaining an inventory - is a challenge for many data centers.

Even though a California system has only a few types of firmware, it would still be a challenge to manage given the fact the system supports hundreds of servers. That is why the UCSM has native firmware management capabilities for all the system components. For server related firmware, these can be driven by policies and driven by a service profile (the firmware follows the service profile, and the blade gets updated with correct version). This helps administrators to perform manual or automatic firmware management of the California system. It also minimizes the risk of any incompatibility between the installed OS drivers and running firmware in adapters and even in some rare cases OS issues that are related to the BIOS version.

6.4.5. Firmware download formats

The California system firmware images can be downloaded into two formats, as a full bundle or a component image. A full bundle contains all the latest firmware for all supported components in a California system (see Figure 143). By downloading this bundle, the administrator knows that these are the latest

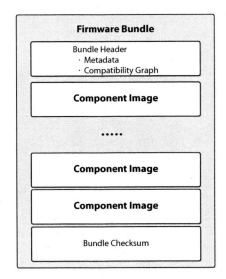

Figure 143: Firmware bundle

versions and have all been tested together and are supported. A component image (see Figure 144) is a part of a bundle. It contains firmware for a single component type in the system, like a BIOS for a server, or a type of adapter. The firmware bundle and component image contain firmware, metadata and checksum. The metadata contains information about the firmware and a compatibility graph. The compatibility graph describes the other firmware in the system that this particular firmware is compatible with. The California sys-

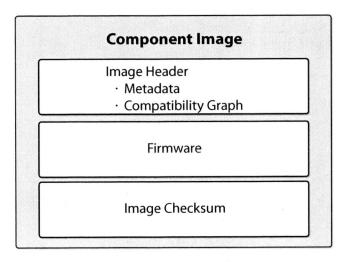

Figure 144: Firmware Components

tem calculates a checksum for the bundle and component image by comparing the calculated checksum against the checksum in the bundle and component image. The California system verifies that the information in the bundle and component image is consistent and not corrupt. Corrupted bundles and component images are discarded and the administrator is notified.

New firmware versions are typically backwards compatible with at least three older versions. If an old and incompatible firmware version is detected in the system, the administrator is alerted that the component needs to be updated. The UCS system does this through a small set of golden API's that never change between versions or during the component and firmware lifetime. Fabric Extenders are automatically upgraded to the latest version if the firmware is too old.

6.4.6. Firmware Life Cycle

All Supported firmware for a California system are downloadable from Cisco in the form of a bundle or component image. The operator downloads the bundle or component image from the Cisco website to a local desktop (single file). The California system downloads the firmware file using any of the following protocols: FTP, TFTP, SCP or SFTP. The firmware file is stored in the fabric interconnect local repository (storage). Firmware that is uploaded into the repository is automatically replicated to a standby UCSM in an HA environment.

The fabric interconnect unpacks the bundle or component, verifies the checksum and reads the header. Based on the metadata in the header, the California system then sorts the firmware based on type and presents it to the administrator as an "installable" image.

The repository is viewed and managed using the definitions in Table 19.

Distributable	All full bundles in the repository
Downloaders	Where and how the California system downloads the firmware.
Firmware Image	A list of all components images in the repository (firmware and header)
Installables	All installable firmware in the system

Table 19: Firmware repository definitions

Table 20 and Table 21 are a description of firmware images in the system.

California system	Description/Purpose
System	This is UCSM, the embedded manager
Fabric Interconnect	Kernel
Fabric interconnect	OS
Fabric Extenders	Chassis controller and interconnect management

Table 20: California firmware images

Server	Description/Purpose
Management Controller	BMC, out-of-band access to blade
BIOS	A library of basic input/output functions
Storage Controller	Local RAID controller for internal disks
Adapters	Management and protocol support
Host NIC	Third-party specific NIC firmware
Host HBA	Third-party specific HBA firmware
Host HBA OptionROM	Third-party specific OptionROM firmware

Table 21: Server firmware images

An installable image is firmware ready to be distributed to a component. Depending on the component, the administrator typically has two ways to instruct the California system to install firmware to one or more components, manually or via a policy (Management Firmware Pack Policy, and Host Firmware Pack Policy). Once initiated, firmware is loaded from the repository down to the component. This is fully automated and controlled by UCSM. No administrative action is needed.

Most components in the system have two firmware banks, one "Running" and one "Backup". The running bank is the current running firmware and the backup bank is the previous version of firmware. Updates to the firmware are always done to the backup firmware bank leaving the running bank untouched. The benefit with this approach is that the backup firmware bank can be updated with different firmware during production without affecting the component itself. There is a boot pointer (startup version) that instructs the component which bank to load during the boot up. The update procedure does not affect this boot pointer so even if the component is repowered it will still be the same firmware as it was running previously. The boot pointer is changed via an "Activate" or "Startup" command. The administrator manages these firmware banks. The administrator can activate or startup a certain available firmware version, the bank containing the firmware version selected by the administrator, will be declared startup bank.

The difference between activate and startup of the firmware is simply that activate will change the boot pointer and then reboot the component in one single command. This will force the component to boot the newly activated firmware. The startup only changes the boot pointer but does not instruct the component to reboot, so the running version will not change.

Figure 145 shows an adapter with its states, including a running and backup version. The picture also shows the bank where a installable firmware would be updated.

Some components do not need two banks. For example, the firmware for the System (UCSM) and the Fabric interconnect (OS and Kernel) does not have the concept of active and backup firmware banks. They do have the concept of a boot pointer though, and the same concept of activation and startup versions. The reason is that the firmware repository which contains all firmware is located in the fabric interconnect internal storage and therefore accessible at boot time to all of these components. They only need a boot pointer to know which firmware to boot from the repository at startup time (see Table 22).

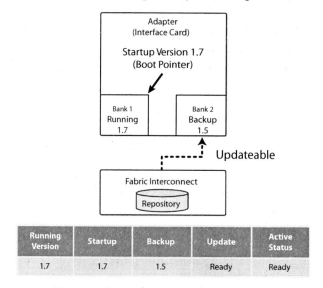

Figure 145: Adapter with Firmware

Firmware Status	Description
Startup Version	This version will be booted at next reboot of the component. (Boot Pointer)
Running Version	This is the currently running version.
Backup Version	This is the previous running version, it is also where the installer will install the next update.

Table 22: Firmware status

Figure 146 shows an example of firmware activation.

6.4.7. Management Firmware Pack Policy

The Management Firmware Pack Policy contains management firmware images for the BMC. A service profile that uses this policy will load the firmware to the BMC on the server associated with that service profile.

6.4.8. Host Firmware Pack Policy

The Host Firmware Pack Policy contains host firmware images, like BIOS, Adapters and local storage controller. A service profile that uses this policy will load all the applicable firmware into the different components of the server associated with the service profile.

A pack can contain multiple firmware for different components. Components that are not defined in the policy are ignored and firmware in the policy that are not applicable, are also ignored.

For example, a host firmware pack can contain multiple firmware for different CNA vendors. When a service profile is associated with a particular server, the firmware pack policy (defined in the service profile), will look to see if an CNA is present in the physical server and if there is a compatible firmware defined in the policy. The CNA will be loaded with that firmware, otherwise, the CNA will keep the current firmware.

6.4.9. The Stateless Computing Deployment Model

All components (except UCSM) in a California system are stateless. This means that the components do not keep a persistent state and can be replaced

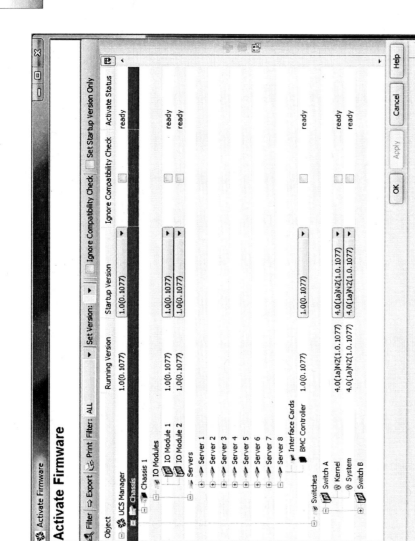

Figure 146: Firmware activation

at any time without losing any configuration information. Since all chassis and servers in the California system are stateless (except for any local disk data), they can be thought of as "Just a Bunch Of Compute nodes" (JBOCs).

The service profile is a logical definition of a server personality and its connectivity. The service profile typically uses policies for defining its behavior and configuration. When a service profile gets associated with a server, the server adopts the personality and connectivity from the service profile. A service profile can be moved from one server to a different server in the same Cisco California system (between multiple California systems an export and import mechanism is available). To take full advantage of the flexibility of statelessness, the local disks on the servers are recommended to only be used for swap or temp space and not to store data since it could limit the mobility of the server.

The personality of the server includes the elements that identify that server and make it unique, see Table 23.

Qualities	Number of Processors, Cores, Memory, type of Adapters, etc (from pool)
BIOS and Firmware	The firmware running on the server and management controller
UUID	Unique identifier of server
MACs	MAC addresses for all defined vNICs in the service profile
WWNs	WWNs for all defined vHBAs in the service profile
WWNN	Node name for server
Boot Settings	Where to boot from, e.g., Virtual Media, Local disks, SAN, and LAN
Local disk configuration	RAID level, for example "mirror", or "best effort mirror" (if local disks are present).
vNICs	Connectivity, VLAN, HA, quality and performance definitions
vHBAs	Connectivity, VSAN, quality and performance definitions
IPMI	IPMI connectivity information

Table 23: Server personality

6.4.10. The Basic Computing Deployment Model

The California system is very powerful due to it stateless nature, however, the system is also able to hide the fact that it is stateless by providing a concept of default service profiles. All configurations in the service profile are optional, except the name of the service profile itself (a.k.a. default service profile). If a default service profile (or an empty service profile) is created and assigned to a server, the server will behave similarly to a traditional server by using pre-allocated IDs from the server itself and default values for connectivity. In this way, there is a 1-to-1 mapping between a physical blade and its identity. Policies can be defined so that the California system can automatically create default service profiles associated with newly discovered servers. That way the California system behaves similarly to a traditional server system.

6.4.11. System Setup - initial setup

Table 24 describes a few configuration questions that need to be answered before the system is ready for use.

Questions	Description
Installation method	Initial installation via GUI or CLI
Setup mode	Restore from backup (Recover) or initial setup
System configuration type	Standalone or cluster configuration
System name	Unique system name, e.g., My_UCS
Admin password	Password for the initial administrator "admin"
Management port IP address and subnet mask	Static IP or DHCP

Table 24: Configuration questions

Once the initial setup is complete, the system is ready for use. From this point all policies and configurations made by the administrator will depend on the deployment model in the data center. Basic deployment is similar to a traditional environment (via default service profiles), or of true stateless deployment (via service profiles). A hybrid implementation is also possible where certain servers are treated like a traditional server, and others are stateless.

Section 6.4.12. and Section 6.4.13. contain two extreme models: the very basic and the stateless model. The California system enables the data center administrator to decide to what extent and for which system they want stateless deployment.

6.4.12. The Default Computing Deployment Model

In this deployment model there is only "the administrator" in the California system. The administrator uses the built in admin account and can perform any operator task as needed. The administrator may set up the basic behavior of system. For example:

- Communication services, (Call Home, Interfaces and protocols disable, enable) Policies;

- Faults, Events, Audit logs, collection retention periods, intervals, Policies, etc.;

- Statistical collection intervals and reporting policies.

Servers are treated as traditional servers and are automatically configured for default service profiles. Each server runs its dedicated software and there is a 1:1 relationship between the OS Instance and the actual server. This means that the servers are using pre-defined values and IDs derived from H/W (MAC addresses, WWNs, UUIDs).

If the server goes down and needs to be replaced (the software might have been protected through a cluster solution), there are probably some infrastructure components affected by the replacement of the server. A network service that depends on a MAC address needs to be reconfigured, SAN-attached storage is typically zoned based on WWN and access layers in the storage device are typically defined based on WWN so this also needs to be reconfigured before the production can start on the OS instance again.

6.4.13. The Stateless Computing Deployment Model

The administrator may change the default configuration system behavior in the following areas:

- Authorization Policies (Local, LDAP, RADIUS, TACACS+);

- Define an organizational structure for resource management;

- Define subject matter expert administrators, for example:
  * Network Administrators;
  * Storage Administrators;
  * Server Administrators;
  * Operations Administrators.

Depending on the customer and the organization, these tasks might be performed by a subject matter expert administrator, but restricted by RBAC in his or her organization.

- Communication services, (Call Home, Interfaces and protocols disable, enable) Policies;

- Faults, Events, Audit logs, collection retention periods, intervals, policies, etc.;

- Statistical collection intervals and reporting policies.

Typically before a California system is ready for full stateless deployments of a server and compute resource, there are policy definitions and rules that need to be configured. The configurations are mainly done upfront and typically involve multiple different administrators. These administrators are experts in their own areas, like Network, Storage, server and application/business, or service administrators.

Leveraging RBAC (users, roles, privileges and possible organizations and locales), the administrators are responsible for setting up the wanted behavior via qualities in their responsibility area.

What follows is an example of responsibilities and tasks that the administrators would set up prior to production. Once set up, the system is pretty much automatically driven by policies and rules.

Network Administration

A network administrator performs California system internal network related tasks by defining policies that describe the required configuration or behavior to connect servers to external network, including:

- Configure an uplink port;

- Configure port channels;

- Configure PIN (PINning) groups (groups containing Ethernet ports);

- Create VLANs;

- Configure the QoS Classes (see Figure 147);

- Configure QoS definitions based on the QoS Classes (see Figure 148);

- Create MAC address pools;

- Create Network related threshold policies for faults, alarms, and statistics;

- Create NIC Adapter profiles.

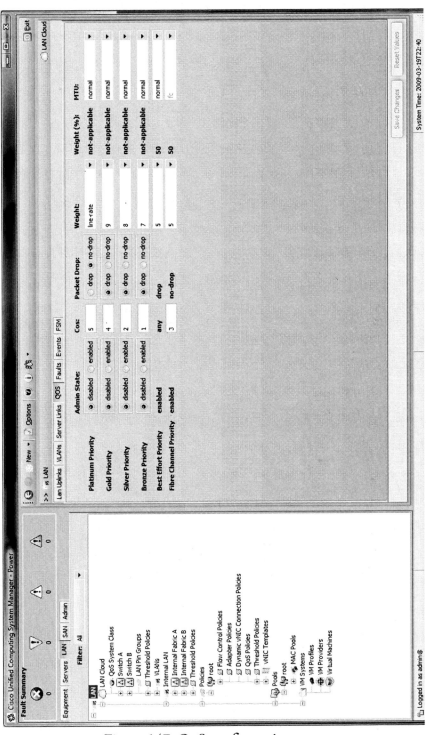

Figure 147: QoS configuration

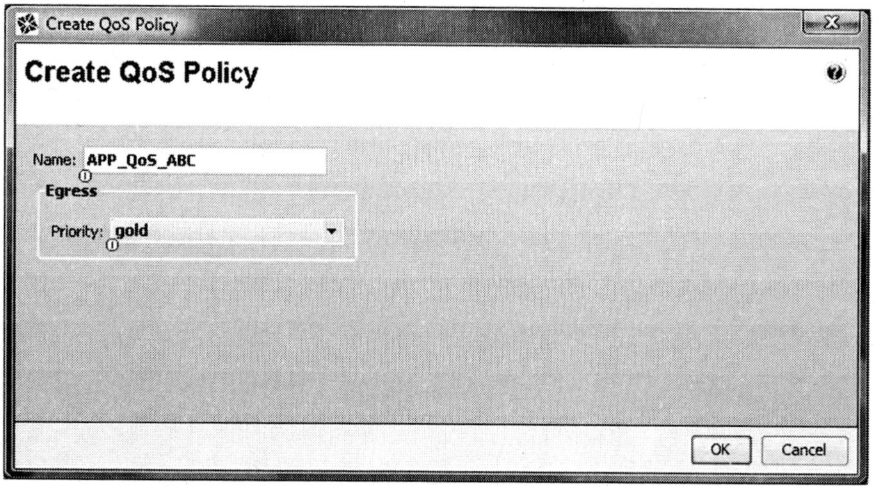

Figure 148: QoS policy definition

Storage Administration

A storage administrator performs California system internal storage related tasks by defining policies that describe the required configuration or behavior to connect servers to an external SAN, including:

- Configure PIN (PINning) groups (groups containing Fibre Channel ports);

- Create VSANs;

- Configure the quality of service classes (QoS Classes);

- Create storage related thresholds, policies for faults, alarms, and statistics;

- Create WWN address pools;

- Create vHBA policies (see Figure 149).

Server Administration

A server administrator performs server management-related tasks by defining policies that describe the required configuration or behavior to instantiate service profiles within the system, including:

- Create server pools;

- Create pool policies and pool qualification policies;

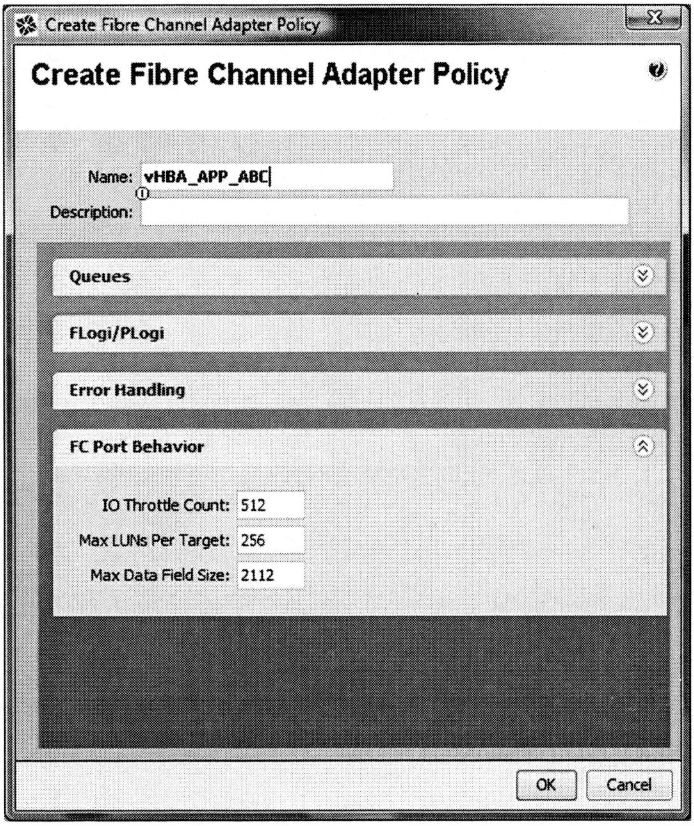

Figure 149: vHBA policies

- Create scrub policies;

- Create IPMI policies;

- Create firmware pack for host and management;

- Create server related threshold policies for faults, alarms, and statistics;

- Create service profile templates.

Production

The normal day-to-day production work is minimal since most components, configurations and system behavior are defined prior to production. Network and Storage administrators rarely need to go in and make changes to policies.

Many policies like, QoS, Adapters, thresholds, and qualifications policies are application dependent. To clarify, an SLA (Service Level Agreement) related to a business service (Application, like HR, Finance, Order Entry, etc) typically dictates the required qualities (HA, performance, portability, etc) for compute resources (servers) that host the business service. These application needs are described in the form of policies. When new applications are introduced into the California system it is recommended that policies describing the needs for the particular application be created. That way the likelihood of an important business service being affected by a less important business service is reduced.

Day-To-Day Server Administration

Day-to-day server management related tasks within the system, include tasks like:

- Create service profiles from pre-created templates;
- Power on/off a server;
- Move a Service Profile between servers;
- Backup the system;
- Monitor logs;
- Firmware maintenance.

6.4.14. Requirements for stateless Service Profiles

The service profile provides a compute service for a software and business services, therefore the creation of a service profile and its qualities typically originate from the requirements of the application. To illustrate this better, let's make up an application "ABC" with reasonable requirements.

It is a business critical application that needs high availability and high performance. It also needs to be able to scale out quickly in cases of unexpected production peaks. The nature of the application in this example is that there is a database that generates a lot of SQL queries and working with large amounts of data. The application administrator understands the behavior of the application well, for example that a particular application is storage I/O intensive and it requires a large amount of memory, but the network traffic load is average.

To break this down, the application has a few requirements:

- High Availability;
- Memory intensive;

- Storage I/O intensive;
- Large amount of storage;
- Average network traffic load;
- Be able to scale out quickly.

In this scenario there might be up to 4 different administrators that together define a service profile, to accommodate all the application requirements.

High Availability

The California system provides multiple levels of HA, from redundant fabric interconnects, I/O paths, power supplies, hardware NIC failover. Table 26 describes different HA components policies, etc.

Memory

The California system supports heterogeneous memory configuration on the servers. If an application or business service requires a minimum amount of memory, this can be defined in a server pool qualification policy, as the one in Table 25.

Minimum RAM	Description	Policy	Admin
Server	The server admin uses the same server qualification policy "APP_HW_Req_ABC", (from the adapter requirement) to add the minimum memory requirements.	Server pool and Qualification policy: "APP_HW_Req_ABC"	Server

Table 25: Example of a server pool qualification policy

Storage I/O intensive

This requirement together with portability and large amount of data, builds the case for SAN attached storage (see Table 27).

Redundancy	Description	Policy	Administrator
Fabric Interconnects	The server administrator needs to confirm that the system in mind is configured for Fabric Interconnect redundancy (clustered).	NA Physical requirement	Server
I/O Paths & Adapters	The server administrator needs to confirm that the system in mind is configured for I/O and Adapter HA.	NA Physical requirement	Server
Power Supplies	The server administrator knows needs to confirm if the system has redundant power supplies.	NA Physical requirement	Server
Hardware NIC failover	There must be at least one VLAN configured in each fabric interconnect, this is done through a VLAN policy. Here the network administrator creates the VLAN Policy "Prod_ABC", which defines dual VLANs. The hardware failover configuration is defined in a policy definition. The vNIC Template is the policy configuration, the network administrator therefore creates the vNIC Policy "APP_vNIC_ABC". Please note that the network administrator will define this policy so it get its MAC address definition from the pool "APP_MAC_PROD"	VLAN: "Prod_VSAN_ABC"* vNIC Template: "APP_vNIC_ABC"	Network

Storage paths, Storage, dual SAN fabrics.		Server and Storage
This is a storage implementation requirement as well as a UCS system requirement.		Server Pool: "APP_SRV_ABC"
The storage administrator needs to confirm that dual fabrics are used, as well as a fault tolerant storage array, and that proper RAID configuration is present on the LUN's.		Server Pool qualification policy: "APP_HW_Req_ABC"
The UCS system need to provide multiple FC adapters for redundant FC paths. The server administrator creates a "server pool qualification policy" ("APP_HW_Req_ABC") to make sure the server will meets the requirements and a "Server Pool policy" ("Populate_ABC_SRV_APP") that defines which pool a qualifying server should be a member of.		Server Pool Policy: "Populate_ABC_APP"
The storage administrator also need to make sure there is at least one vSAN defined in each fabric interconnect to provide dual fabric connectivity. This is done through a vSAN policy ("Prod_VSAN_ABC").		VSAN: "Prod_VSAN_ABC"***

OS and Application		NA	Server and Application
The server administrator and the application administrator will need to make sure the proper drivers are installed in the OS, and that the application is configured correctly.			
Server mobility via service profiles, shorten time to fix, and maintenance windows.	If the server goes down and needs to be replaced, the service profile should be able to quickly be started on a different server. The server administrator creates a UUID pool, to make sure the service profile can receive a portable UUID ("APP_UUID_PROD"). The Network administrator creates a MAC pool ("APP_MAC_PROD"), to make sure the service profile can receive portable MAC addresses. The Storage Administrator creates a WWN pool ("APP_WWN_PROD"), to make sure the service profile can receive portable WWN addresses.	UUID Pool: "APP_UUID_PROD"*** MAC Pool: "APP_MAC_PROD"*** WWN Pool "APP_WWN_PROD"***	Server Network Storage

*Please note that multiple VLAN policies can point to the same VLAN ID.

**Please note that multiple VSAN policies can point to the same VSAN ID.

***Please note that overlapping pools are allowed, pools can overlap other pools with same ID's and it is possible to create a separate pool for each application if desired.

Table 26: Example of HA component policies

Component	Description	Policy	Admin
vHBA	The storage administrator creates vHBA template to describe the quality requirement for the vHBA ("APP_vHBA_ABC"). Please note that the storage admin defined this policy so it get its WWN definition from the pool "APP_WWN_PROD"	vHBA Template: "APP_vHBA_ABC"	Storage

Table 27: Example of storage policy

Average network load

Even though the network load requirement was average, given the importance of the application, there is a need to make sure that this application has higher priority than regular applications. This may come in handy if there is an ill behaved or suddenly a non-important application starts to utilize the network heavily (see Table 28).

Component	Description	Policy	Admin
vNIC	The network administrator creates a QoS of priority class gold (Second highest priority), this ensure that any application in Silver, Bronze, or best effort classes will not affect the application ("APP_QoS_ABC").	QoS Policy: "APP_QoS_ABC"	Network

Table 28: Example of network policy

Being able to scale out quickly

Since all of the definitions are policies, service profiles and service templates can leverage them. A "Service Profile Template" is a natural choice to meet this requirement of scale out of server resources. A service profile template is similar to a service profile, with the exception that it cannot be associated with a server. It is used for instantiating new service profiles. By using a Template any administrator can easily create a compute resource that meets the requirement of the business service (see Table 29).

Component	Description	Policy	Admin
Service Profile Template	The server administrator creates the service profile template, and defines all the policies that has been created by the various administrators for the application ABC ("APP_ABC")	Service Profile Template: "APP_ABC"	Server

Table 29: Example of a service profile template

No pre-defined policies

Table 30 might be an extreme scenario with no pre-defined policies.

Policy/Template	Name
MAC Pool	APP_MAC_PROD
vHBA Template	APP_vHBA_ABC
vNIC Template	APP_vNIC_ABC
QoS Policy	APP_QoS_ABC
Server Pool	APP_SRV_ABC
Server Pool qualification policy	APP_HW_Req_ABC
Server Pool Policy	Populate_ABC_APP
UUID Pool	APP_UUID_PROD
VLAN Policy	Prod_VSAN_ABC
VSAN Policy	Prod_VSAN_ABC
WWN Pool	APP_WWN_PROD
Service Profile Template	APP_ABC

Table 30: Examples of policies and templates

The following policies and templates have been created, by the administrators, based on the application requirement This is just to illustrate how a service profile gets defined, and some of the different policies it can include. As described earlier the California system allows a service profile to be created with minimum configuration, and no policies.

Figure 150 illustrates how these definitions are related and connected to make up the service profile. This picture shows how the vHBA and vNIC templates are consuming other policies and how the server pools get automatically populated via a server pool qualification and a server pool policy.

Figure 151 shows how policies are related with the service profile template, and then instantiated to multiple service profiles.

Figure 152 shows a vNIC template.

6.4.15. System Logging

The California system logs all events, faults and Audit information in the system. These logs can be exported to an external system. Events can also be subscribed to. This is useful for monitoring tools that integrate with the California system.

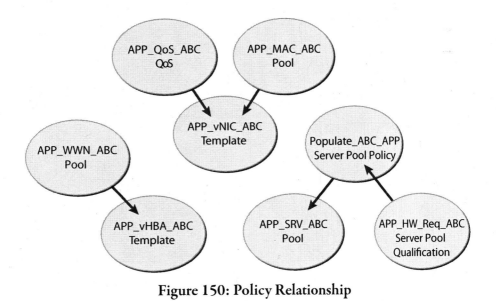

Figure 150: Policy Relationship

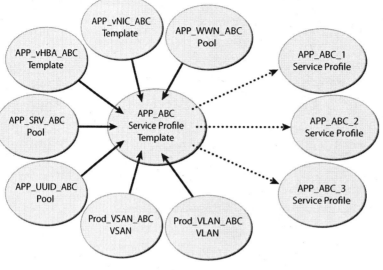

Figure 151: Policies in templates

Figure 152: vNIC template

6.4.16. Faults and Events

UCS Manager exploits generic object notifications in the management of events and faults. An event is a representation of something that momentarily occurred in the system. A fault represents something that failed in the system (or that got created by the triggering of an alarm threshold). Events and faults are both objects that are managed by UCS Manager and are subject to the same base set of rules as other MOs. However, events and faults have additional rules that specifically apply to them.

Events

Events are immutable as managed objects because they correspond to non-persistent conditions in the system. Once an event is created and logged, it does not change. For example, the beginning and ending states of a blade power-on request are logged as events, indicating that UCS Manager has started and completed the request (see Figure 153).

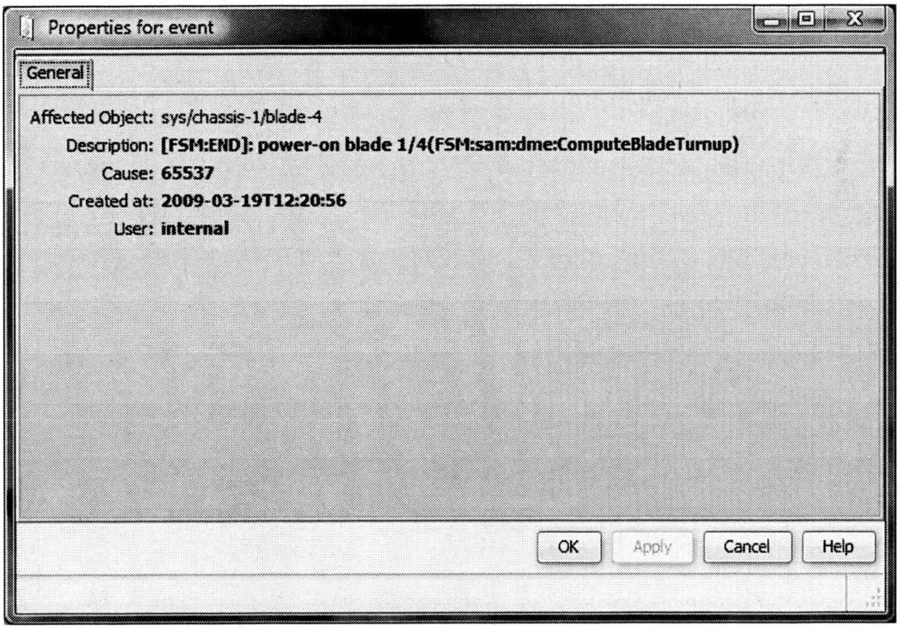

Figure 153: Power-on

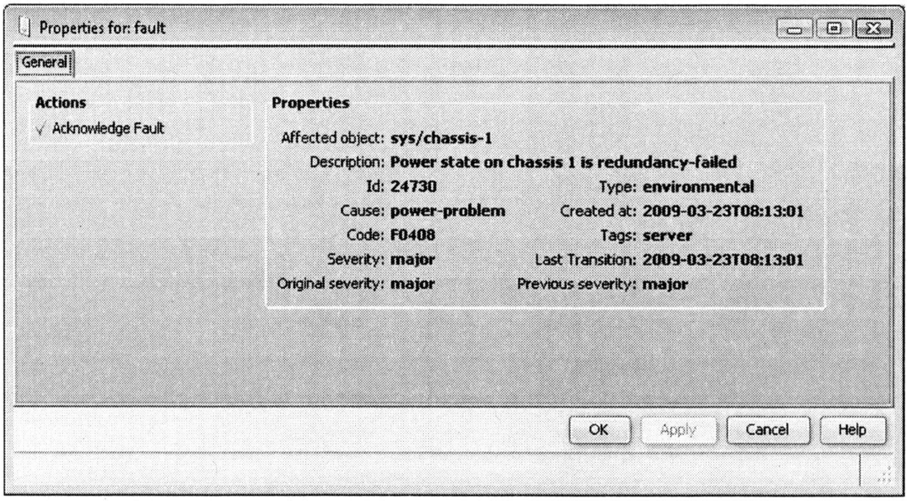

Figure 154: Power problem

Faults

Fault MOs, on the other hand, are mutable because the operational state of the faulted endpoint may, at any time, transition to a functioning state. A corrected fault is known as a cleared fault. For example, physically unplugging a link between a fabric extender and a switch will put the switch port into a faulted state. Plugging the link back in will clear this fault (see Figure 154).

UCS Manager has a global fault policy that defines how to handle cleared faults (whether to automatically delete them or retain them for a certain amount of time).

Statistics

UCS Manager collects various device-specific statistics from the managed endpoints including, adapters, blades, chassis, hosts, and ports. These statistics reflect the operational state of the endpoints. For example, an adapter maintains counters for the number of packets it sends and receives. Other statistics include chassis fan speeds, CPU temperatures, and power consumption.

UCS manager has global statistics policies that define how often statistics are collected from each of the endpoints and reported to UCS Manager.

Thresholds

A threshold policy in UCS Manager defines values for what is considered "normal", "above normal ", and "below normal" for any of the statistics that have been collected. The policy also describes the severity (e.g. warning or critical) of the deviation the from normal value. When a statistic reaches an above-normal or below-normal value (known as an escalating value), UCS Manager will create a fault MO with the corresponding severity. When the statistic value crosses a de-escalating value, UCS Manager will clear the fault.

6.4.17. Audit log

The UCSM audits both direct and indirect user actions in the same way as it audits events. User actions are representative of an event that occurred "momentarily" in the system. These events can be based on a user action such as a creation, deletion or modification of an object. As with all other objects in the system they are exportable (see Figure 155).

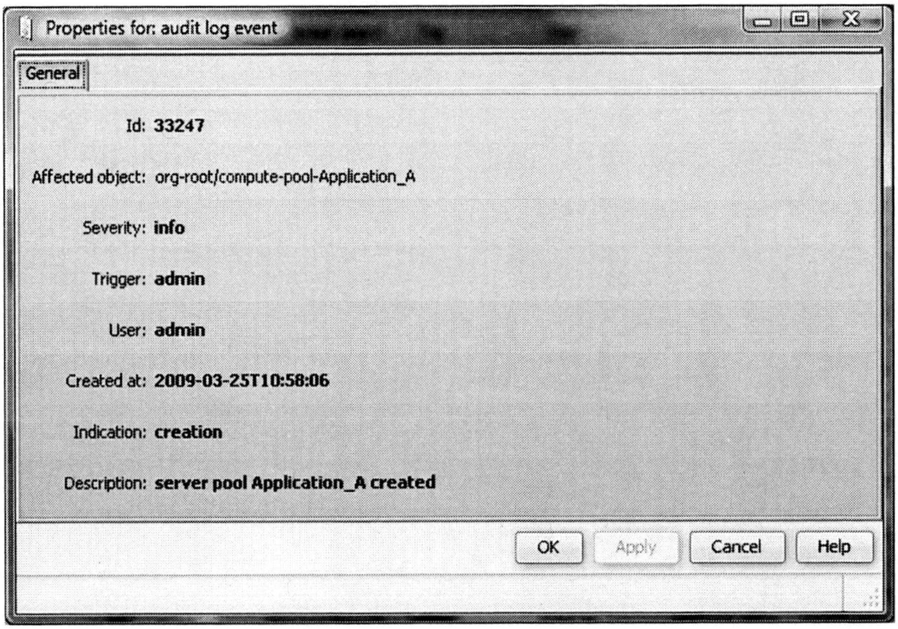

Figure 155: Audit log entry

6.5. BMC® BladeLogic® Integration with Cisco UCS

BladeLogic and Cisco are working together to integrate and add support for UCS in BladeLogic. By leveraging the unique features of UCS, like the policies and service profiles, this integration provides a solution that brings additional value to customers. It gives BladeLogic the ability to manage multiple California systems and to share logical artifacts among them. The integration allows administrators to manage and to deploy servers, and to perform compliance operations (snap-audit and rule-based) on UCS assets. This solution allows the server administrator to use BladeLogic for basic day-to-day UCS system management without using the UCS GUI or CLI.

6.5.1. Integration Point

All UCS instances will be managed from a single agent that can run on any server. This agent has custom objects to manage the UCS environment. The administrator uses the BladeLogic's live view pane to add instances of UCS to the agent. BladeLogic discovers and imports UCS objects using the XML API. These objects include general inventory, profiles, templates, relationships, topology, connectivity, and policy definitions.

6.5.2. Service Profiles and Pools

BladeLogic creates and maintains its own definitions of service profiles and templates. When an administrator uses BladeLogic to associate a service profile with a blade, BladeLogic will automatically create an identical service profile in the local UCS system. If the service profile is disassociated from the blade, BladeLogic will remove the service profile from the local UCS.

ID pools (which include UUID, MAC, and WWN pools) are also locally created and maintained from within BladeLogic. By owning the service profile definitions and the identities, BladeLogic can manage multiple UCS systems and provide service profile mobility across UCS systems.

Blade pools are created and maintained within UCS and are consumed by BladeLogic.

6.5.3. Policies

The policies are locally maintained and managed from within the UCS, but they are consumed by the service profiles defined within BladeLogic. The policies are owned by the server, storage and network administrators in the local UCS. This maintains a clean administrative structure where each policy is defined by a local administrator.

6.5.4. UCS Actions from BladeLogic

Many operator actions have been integrated and can now be performed from BladeLogic, such as:

- Organization – create, delete, modify, browse;
- ID pool – create, delete, modify, browse;
- Service profile – create, delete, associate, dissociate, move, modify, browse, and power actions (e.g. start and stop);
- Launch KVM (keyboard/video/mouse) to server;
- Service profile template - create, delete, modify, browse;
- RBAC –handled by BladeLogic;
- SAN boot definitions.

This integration enables an administrator to perform end-to-end provisioning, including:

- Service profile creation (with all features from UCS);
- Service profile association;
- Service profile startup;
- OS install (local or SAN);
- OS boot;
- Application installation;
- Patch installation.

7. Planning a California installation

Good initial planning is the first task of a successful California installation in a Data Center. To create a good plan for installation one must look into power and cooling capabilities, cabling, bandwidth requirements, desired availability levels and much more. In this chapter we take a closer look on how to plan for a successful California installation.

7.1. Power and Cooling

A California system has two main components that require connection to an external power source – the UCS 6100 Fabric Interconnects and the UCS 5108 Blade Server Chassis. Both of these components use 220V AC single-phase electric power and can be configured for different modes of power redundancy. In case 220V AC single-phase electric power is not readily available, Power Distribution Units (PDUs) are used to convert electric power from three phases to a single phase.

The two models of Fabric Interconnects have different power consumption characteristics. Their values are listed in Table 31.

	UCS 6120XP 20 Port Fabric Interconnect	UCS 6140XP 40 Port Fabric Interconnect
Normal Power Consumption	350W	480W
Maximum Power Consumption	450W	750W

Table 31: Fabric Interconnect power consumption

Each Blade Server Chassis can hold four power supplies that are rated at 2500W each. Power supplies can be replaced from the front of the chassis as shown in Figure 156.

Normal power consumption for a Blade Server Chassis depends on the blade server configuration (memory, I/O Ports, etc) and the load on the CPU imposed by the applications being run. Many of today's non-virtualized servers run at around 20 - 30% utilization.

Figure 156: Front of the chassis and power supply

Power consumption numbers are indicative only since they are based upon preliminary measurements. Some of the numbers measured on a fully loaded chassis with eight blades are reported in the Table 32[1]. Each server blade is configured with two Intel® X5570 CPUs, 6 DIMMs, a single mezzanine card and single hard disk.

CPU load	Combined power Consumption for chassis and 8 blades
100%	2.6KW
50%	2KW
Idle	1KW

Table 32: Preliminary power consumption

Power consumption is highly dependent on the blades configuration and workloads and therefore these numbers represent a very limited sample. Workload that was used to create load in this test was a standards based Java benchmark.

All power supplies in a California system are rated at 90% power efficiency or better and are hot swappable.

Fabric Interconnects and Blade Server Chassis are both air cooled with front to back airflow to make them suitable in a typical data center cooling environ-

---

1   These numbers should not be used as normal power consumption figures. Please consult the appropriate documentation.

ment. They can support commonly used hot isle/cold isle cooling strategy.

The Blade Server Chassis has eight large fan modules in the back that are hot-swappable, as shown in Figure 157.

Power and cooling capabilities in the data center are the primary considerations that determine how many California Blade Server Chassis can be installed per rack. This combined with the number of blades required provides the number of racks that are needed for the installation.

To achieve accurate cooling requirements for a California system, it is necessary to know its power consumption. That number can then be translated into different cooling values. There is a worldwide trend among standard-setting organizations to move all power and cooling capacity measurements to a common standard, the Watt. In North America "British Thermal Units" (BTUs) and "Tons" are still commonly used to describe cooling requirements. Table 33 can be helpful when translating energy consumption into BTUs and Tons.

From KWs to	Multiply KWs by
BTUs per hour	3414
Tons	0.284

Table 33: KWs, BTUs, and Tons

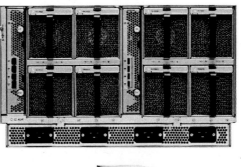

Figure 157: Back of the chassis and fan modules

7.2. Physical sizing and Environmental requirements

A California system is designed to fit into a standard 19-inch rack. Different components differ in height, depth and weight, but their width remains the same. The following table describes sizes for different elements.

Description	Specification
Height	1.72" (4.4cm) - 1 RU
Width	17.3" (43.9cm)
Depth	30" (76.2cm)
Weight	
With 2PS, 1 Expansion and 2 Fan modules	35 lb (15.88 Kg)
Temperature, operating	32°F to 104° F (0 to 40° C)
Temperature, non-operating	-40 to 158° F (-40 to 70° C)
Humidity (RH), non-condensing	5 to 95%
Altitude	0 to 10,000 ft (0 to 3,000 m)

Table 34: 20 ports Fabric Interconnect

Description	Specification
Height	3.47" (8.8cm) 2 RUs
Width	17.3" (43.9cm)
Depth	30" (76.2cm)
Weight	
With 2PS, 1 Expansion and 2 Fan modules	50 lb (22.68 Kg)
Temperature, operating	32°F to 104°F (0 to 40° C)
Temperature, non-operating	-40 to 158°F (-40 to 70° C)
Humidity (RH), non-condensing	5 to 95%
Altitude	0 to 10,000 ft (0 to 3,000 m)

Table 35: 40 ports Fabric Interconnect

Description	Specification
Height	10.5" (26.7cm) - 6RUs
Width	17.5" (44.5cm)
Depth	32" (81.2cm)
Weight	250 lb (113.39 Kg) fully loaded
Temperature, operating within altitude: 0 - 10k feet (0 - 3,000 m)	50 to 95° F (10 to 35° C)
Temperature, non-operating within altitude: 0 - 40k feet (0 - 12,000 m)	-40 to 149° F (-40 to 65° C)
Humidity (RH), non-condensing	5 to 93%
Altitude	0 to 10,000 ft (0 to 3,000 m)

Table 36: Compute chassis

7.3. Connectivity

Fabric Interconnects provide connectivity for management and data traffic. All I/O required by a California system goes through the Fabric Interconnects. The two models of Fabric Interconnects provide either 20 or 40 fixed 10GE ports. Additionally there are one or two expansion slots respectively. These slots can be used to add more connectivity options if needed (see Section 5.1.2.).

Each port on the Fabric Interconnects is configured as an access or as an uplink port. Server ports are connected to the Fabric Extenders that reside in the Blade Server Chassis.

The Fabric Extenders provide connectivity from the Blade Server Chassis to the Fabric Interconnect (see Figure 158). Each Fabric Extender has 8 internal server-facing 10GE ports and 4 uplink ports connecting to the Fabric Interconnects.

Different California blades have the capability of housing either one or two mezzanine cards. All mezzanine cards are equipped with two 10GE ports. These two ports are connected to the Fabric Extenders via the chassis midplane. The left port of each mezzanine goes to the left Fabric Extender and the right port of each mezzanine goes to the right Fabric Extender.

The upstream links of the Fabric Interconnects can work either in switch mode or end-host mode. The difference between the two modes is that in switch

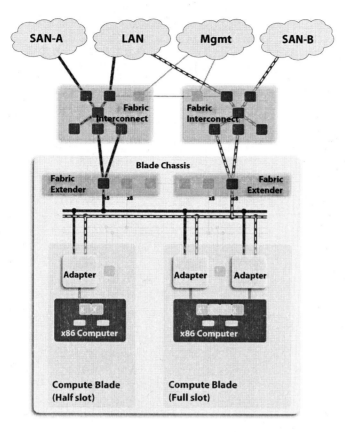

Figure 158: Connectivity

mode the Fabric Interconnect participates in the Spanning Tree protocol and in end-host mode the Fabric Interconnect does not participate in the Spanning Tree protocol and it looks like a host.

Connectivity mode to SAN networks is always done in end-host mode (aka NPV: N_Port virtualization). In this mode, all California FC (Fibre Channel) ports present themselves as a host ports or N_Ports. This allows a California system to connect to any vendor SAN without using special interoperability modes. Moreover, this mode does not consume any FC domain.

7.4. Choosing the right cables

Each California Fabric Extender can be connected to the Fabric Interconnects either with Twinax or fiber cables. There are important details that need to be considered when deciding which type of cable to use for connectivity. One of the most important factors is the number of racks the California system covers. That determines the minimum cable length required to connect all California Blade Server Chassis to the Fabric Interconnects.

The California system uses standard SFP+ modules for all required connectivity.

7.4.1. Twinax

Twinax is the most cost effective solution to connect Fabric Extenders to the Fabric Interconnects, but it has limitations in length. The maximum length of Twinax cable for 10GE is 7m (23 ft) and that will not be enough for all situations particularly in a multi-rack installation.

Power consumption of Twinax connection is very minimal and is currently rated at ~0.1W per cable.

7.4.2. Fiber

Fiber cabling should be used if the distance between Fabric Interconnects and Fabric Extenders is too long for the Twinax cable. These situations could occur when either creating large California installations that span over multiple racks and/or if "end of the row" network design is used.

7.5. Bandwidth

The size of a California system is dependent on bandwidth requirements. Required bandwidth dictates how many uplinks are needed from each Blade Server Chassis to each Fabric Interconnect.

Fabric Extenders and Fabric Interconnects can have one, two or four 10GE links between them. Table 37 summarizes available bandwidth for a Blade Server Chassis and for each individual server slot, if they are evenly loaded.

It is important to notice that the smaller number is per mezzanine card and not per blade. As new blade models come to market they may contain more than one mezzanine card. Also, it is equally important to understand that this value is normalized and it assumes that all mezzanine cards are simultaneously sending as much traffic as possible. All cards share unused bandwidth.

The number of links used to connect the Fabric Extenders and the Fabric Interconnects limits the size of a single California system. Table 38 summarizes the maximum size of a California system with different numbers of uplinks and with the different Fabric Interconnects models.

Number of uplinks from each Fabric Extender to each Fabric Interconnect	1	2	4
Available bandwidth for Blade server Chassis with one Fabric extender	10 Gbps	20 Gbps	40 Gbps
Available bandwidth for each mezzanine card with one Fabric Extender	1.25 Gbps	2.5 Gbps	5 Gbps
Available bandwidth for Blade Server Chassis with two Fabric Extenders	20 Gbps	40 Gbps	80 Gbps
Available bandwidth for each mezzanine card with two Fabric Extenders	2.5 Gbps	5 Gbps	10 Gbps

Table 37: Planning for bandwidth

Number of uplinks from each Fabric Extender to each Fabric Interconnect	1	2	4
Maximum number of Blade Server Chassis with UCS 6120XP 20 Port Fabric Interconnect	20	10	5
Maximum number of Blade Server Chassis with UCS 6140XP 40 Port Fabric Interconnect	40	20	10

Table 38: Planning for blade server chassis

7.6. Planning for redundancy

When planning for redundancy one must look at the big picture. Redundancy, in its simplest form, is avoiding single points of failures. Take for example power supplies. When powering normal 1RU or 2RU rack-mounted servers it is possible to survive with one power supply, but two are needed for redundancy, in case the first happens to fail. The same rules apply to California as well.

7.6.1. Power supply redundancy

California supports multiple levels of redundancy in its power configuration. A California system can be configured with either N+1, N+N and grid redundant configurations. A single Blade Server Chassis can survive under any load and in any configuration with just two power supplies active. Table 39 illustrates different levels of power redundancies in a California Blade Server Chassis that houses eight standard configuration blades. Standard configuration includes dual processors, dual disks, one mezzanine card and twelve DIMMs.

Once connected, all power supplies in a California Blade Server Chassis become active and share the load among them. If one or more power supplies fail during operation, the load from failed power supplies is shared evenly to remaining power supplies.

Number of Power Supplies	Total Face value in KW	Redundancy achieved
1	2.5	Insufficient Power Available
2	5	None
3	7.5	N+1
4	10	N+N / Grid Redundancy

Table 39: Power supply redundancy

7.6.2. I/O redundancy

In today's data centers I/O redundancy is an increasingly important aspect to plan for. Connecting hosts to both Fabric Interconnects duplicates all application I/O paths and eliminates single points of failures.

California I/O capabilities have been designed for maximum redundancy. New technologies have been used to increase I/O availability without the introduction of added complexity. When implementing a California system with two Fabric Interconnects, I/O redundancy is achieved by creating dual fabrics automatically. These two fabrics are separated from each other with respect to I/O, failure domain, and a management perspective.

7.6.3. Ethernet Interfaces redundancy

Today Ethernet redundancy primarily depends on Operating System drivers such as Bonding and NIC teaming. To use these drivers they need to be configured correctly and the switch interfaces must also be configured as an EtherChannel or a similar protocol. This must be done for each node individually which can be an error-prone and time-consuming operation, since multiple operational teams need to coordinate to make this happen.

In a California system, if so desired, this operating model can still be used, but California also introduces a new and improved way of implementing Ethernet redundancy far less error-prone and labor intensive.

When creating Ethernet interfaces in a Service Profile there is an option to create them as "highly available interfaces"[2]. In creating Ethernet interfaces a primary Fabric Interconnect must be selected which determines where that particular interface is bound. If the high availability option is also selected, and there is a failure in the I/O path, this interface is automatically moved from one physical interface to another. This action is seamless and it happens without the operating system being aware of it. It is fully operating system agnostic.

To ensure that this California feature works seamlessly, both Fabric Interconnects must have access to all required VLANs. Once this is verified, any node in the California system can be easily configured for Ethernet redundancy by checking the high availability option. There is no need to install and configure any additional drivers in the operating systems. There is no need to change the upstream Fabric Interconnects as the configuration is programmed at the time of initial deployment. Fabric Interconnects configuration needs to be revisited only if new VLANs or uplinks are connected to the Fabric Interconnect, but in most cases even this does not require major changes to the Fabric Interconnect configuration. This it is a non-intrusive operation that doesn't affect production traffic that is being processed. Once these changes are done to the Fabric Interconnects they are available for all nodes in the California system.

Working methods of this functionality are very simple. When creating Ethernet interfaces for Service Profiles each interface has a primary fabric interconnect that it is connected to. If the interface is configured for high availability it has the permission to move from one Fabric Interconnect to another, in case of a failure. Once the failure condition is cleared, interfaces that were affected

---

2   Check the appropriate documentation as this option is supported on most but not all adapters.

will automatically move back to their primary Fabric Interconnect.

This new method for creating redundant Ethernet interfaces provides multiple benefits. One obvious benefit is the simple configuration, which not only saves time but reduces the possibility of configuration errors. Other not so obvious benefits range from the reduced number of interfaces required for creating highly available Ethernet connections to the agnostic operating system. This functionality can be used in any Operating system that can be run on a California system, since it does not require separate interface driver. This also removes the possibility of human errors in redundancy configuration and testing with the specific OS patch levels. If a service-profile requires high availability, then any blade that runs that profile will have high availability. This is completely dynamic. A blade that in the morning runs a non-HA Service Profile can in the afternoon run an HA profile.

7.6.4. Fibre Channel Interfaces redundancy

A California system has two distinct fabrics when configured for redundancy. This implies a different high availability model than a typical LAN.

Storage redundancy can be achieved by creating a vHBA (virtual HBA) per fabric. Using a storage multipathing driver then combines the vHBAs (see Figure 159). These drivers are available inside of operating systems and from storage vendors.

Both fabrics in California are simultaneously active and therefore active/active configurations are possible if supported by the storage system. This gives California the capability to load balance storage traffic across different fabrics and to achieve higher bandwidth for storage devices.

Because these multipathing drivers differ from one another, the appropriate documentation should be referenced before implementing these drivers into a California system.

The diagram below shows the placement of a multipathing driver in the OS stack and the communications which take place between the different modules.

When configuring California for storage multipathing a minimum of two vHBAs must be created in the Service Profile. Each vHBA has a virtual World Wide Name (vWWN) that can be used for fibre channel zoning and LUN (Logical Units) masking in storage arrays.

If WWN-based zoning is used, these vWWNs must be members of the ap-

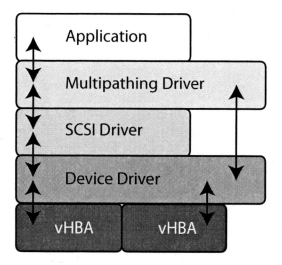

Figure 159: Multipathing Driver

propriate zones. If port-based zoning is used, all possible uplink ports where particular vWWNs may get pinned must also belong to the right zones.

Once all zoning information is in place, hosts and storage arrays can log into the FC fabric where it is possible to verify through management if the vWWNs are logged correctly.

The next step is to map the appropriate LUNs to vHBAs by using their vWWNs. LUN masking is performed in the storage array. Operation is very array-specific so further information on how this is performed should be researched in the relevant storage array manuals.

Once both vHBAs have access to the necessary storage, multipathing drivers are able to configure themselves and create the appropriate device mappings in to the OS layer.

8. Bibliography

8.1. PCI Express

[1] "How PCI Express Works," howstuffworks, a Discovery Company, http://computer.howstuffworks.com/pci-express.htm

[2] "PCI Express," from Wikipedia, the free encyclopedia, http://en.wikipedia.org/wiki/Pci_express

[3] "PCI-SIG Announces PCI Express 3.0 Bit Rate For Products In 2010 And Beyond," OCI SIG newsroom, http://www.pcisig.com/news_room/08_08_07/

8.2. IEEE 802.3

[4] "IEEE 802.3: LAN/MAN CSMA/CD Access Method," IEEE Standard Association, http://standards.ieee.org/getieee802/802.3.html

8.3. Improvements to Ethernet

[5] "Ethernet Enhancements Supporting I/O Consolidation," Nuova Systems, http://www.nuovasystems.com/EthernetEnhancements-Final.pdf

[6] SFF-8431 – "Enhanced 8.5 and 10 Gigabit Small Form Factor Pluggable Module," ftp://ftp.seagate.com/sff/SFF-8431.PDF

[7] "Cut-Through and Store-and-Forward Ethernet Switching for Low-Latency Environments," Cisco Systems, http://www.cisco.com/en/US/prod/collateral/switches/ps9441/ps9670/white_paper_c11-465436.html

8.4. IEEE 802.1 activities

[8] "802.1Qbb - Priority-based Flow Control," IEEE Standard for Local and Metropolitan Area Networks---Virtual Bridged Local Area Networks - Amendment: Priority-based Flow Control." http://www.ieee802.org/1/pages/802.1bb.html

[9] "Enabling Block Storage over Ethernet: The Case for Per Priority Pause," http://www.ieee802.org/1/files/public/docs2007/new-cm-pelissier-enabling-block-storage-0705-v01.pdf

[10] "Fabric Convergence from a Storage Perspective," http://www.ieee802.org/1/files/public/docs2007/au-ko-fabric-convergence-0507.pdf

[11] "802.1Qau - Congestion Notification," IEEE Standard for Local and Metropolitan Area Networks---Virtual Bridged Local Area Networks -

Amendment: 10: Congestion Notification, http://www.ieee802.org/1/
pages/802.1au.html

[12] "802.1Qaz - Enhanced Transmission Selection," IEEE Standard for Local and Metropolitan Area Networks---Virtual Bridged Local Area Networks - Amendment: Enhanced Transmission Selection, http://www.
ieee802.org/1/pages/802.1az.html

8.5.   FCoE

[13] "FCoE (Fibre Channel over Ethernet)," http://www.fcoe.com/

[14] "INCITS Technical Committee T11," http://www.t11.org/

[15] "Open-FCoE," http://www.open-fcoe.org/

[16] "Fibre Channel over Ethernet in the Data Center: An Introduction,"
FCIA: Fibre Channen Industry Association, http://www.fibrechannel.
org/OVERVIEW/FCIA_SNW_FCoE_WP_Final.pdf

[17] "I/O consolidation in the Data Center," Claudio DeSanti, Silvano Gai,
Cisco Press, 2009

8.6.   TRILL

[18] "Transparent Interconnection of Lots of Links (trill)," IETF WG, http://
www.ietf.org/html.charters/trill-charter.html

8.7.   Virtualization

[19] "VMware ESX Server 2.1: Setting the MAC Address Manually for a Virtual Machine, http://www.vmware.com/support/esx21/doc/esx21ad-
min_MACaddress.html

[20] "The Role of Memory in VMware ESX Server 3," VMware® and Kings-
ton® whitepaper, http://www.vmware.com/pdf/esx3_memory.pdf

8.8.   Memory Subsystem

[21] "JESD100B.01: Terms, Definitions, and Letter Symbols for Microcom-
puters, Microprocessors, and Memory Integrated Circuits," JDEC standard, http://www.jedec.org/download/search/JESD100B01.pdf

[22] "JEDEC Standard No. 21–C," JDEC Standard, http://www.jedec.org/
download/search/3_07_01R5.pdf

[23] "Future Vision of Memory Modules for DRAM," Bill Gervasi, VP
DRAM Technology SimpleTech http://www.jedex.org/images/pdf/b_

gervasi_modules.pdf

[24] "IBM® Chipkill® Memory," IBM® Corporation, http://www.ece.umd.edu/courses/enee759h.S2003/references/chipkill_white_paper.pdf

[25] Intel® E7500 Chipset MCH Intel®x4 Single Device Data Correction (x4 SDDC) Implementation and Validation Application Note (AP-726)," Intel®, http://www.intel.com/Assets/PDF/appnote/292274.pdf

[26] "Chipkill correct memory architecture," Dell®, http://www.ece.umd.edu/courses/enee759h.S2003/references/chipkill.pdf

[27] "Ultimate Memory Guide,: How Much Memory Do You Need?" Kingston®, http://www.kingston.com/tools/umg/umg01b.asp

[28] Ultimate Memory Guide,: Different kinds of memory?" Kingston®, http://www.kingston.com/tools/umg/umg05a.asp

[29] "Server Memory" Kingston®, http://www.kingston.com/branded/server_memory.asp

[30] "Memory technology evolution: an overview of system memory technologies," HP®, http://h20000.www2.hp.com/bc/docs/support/SupportManual/c00256987/c00256987.pdf

[31] "FAQ on the NUMA architecture," sourceforge.net, http://lse.sourceforge.net/numa/faq/

8.9. Intel® processors

[32] "Next Generation Intel Core Microarchitecture (Nehalem) Processors," Rajesh Kumar: Intel Fellow http://intelstudios.edgesuite.net/fall_idf/tchs001/msm.htm

[33] "Intel® Core™ i7 Processor The best desktop processors on the planet," Intel®, http://www.intel.com/products/processor/corei7/index.htm

[34] "Intel Launches Fastest Processor on the Planet," Intel®, http://www.intel.com/pressroom/archive/releases/20081117comp_sm.htm

8.10. Data Centers

[35] "Datacenter Power Trends," NANOG 42 Power Panel, http://www.nanog.org/mtg-0802/presentations/Snowhorn-Power.pdf

[36] Virtualization Changes Virtually Everything," Gartner Special Report, March 28, 2008.

[37] Virtualization Evolves into Disaster Recovery Tool," eWeek, May 7, 2008.

[38] Cayton, Ken, "Choosing the Right Hardware for Server Virtualization,"

IDC White Paper, April 2008.

[39] IDC MultiClient Study.

8.11. Green

[40] "ENERGY STSAR," U.S. Environmental Protection Agency and the U.S. Department of Energy, http://www.energystar.gov

[41] "Climate Savers smart computing," Climate Savers® WWF, http://www.climatesaverscomputing.org/

[42] "Average Retail Price of Electricity to Ultimate Customers by End-Use Sector, by State," Energy Information Administration, http://www.eia.doe.gov/cneaf/electricity/epm/table5_6_a.html

8.12. Cloud Computing

[43] "Above the Clouds: A Berkeley View of Cloud Computing," Technical Report No. UCB/EECS-2009-28, Electrical Engineering and Computer Sciences University of California at Berkeley, http://www.eecs.berkeley.edu/Pubs/TechRpts/2009/EECS-2009-28.pdf

[44] "Above the Clouds, A Berkeley View of Cloud Computing," Blog at http://berkeleyclouds.blogspot.com/

9. Glossary, Tables and Index

9.1. Glossary

- 10GBASE-T: A standard for 10 Gigabit Ethernet over twisted pair.

- 10GE: 10 Gigabit Ethernet, see also IEEE 802.3.

- 802.1: An IEEE standard for LAN Bridging & Management.

- 802.1Q: An IEEE standard for bridges, VLANs, STP, priorities.

- 802.1Qbb - Priority-based Flow Control.

- 802.1Qau - Congestion Notification.

- 802.1Qaz - Enhanced Transmission Selection.

- 802.3: The Ethernet standard.

- ACL: Access Control List, a filtering mechanism implemented by switches and routers.

- Aka: Also Known As.

- AG (Application Gateway): Management process that focuses on monitoring and configuring a single elements of the system such as the fabric extender, the server, the I/O adapter, etc.

- AQM: Active Queue Management, a traffic management technique.

- B2B: Buffer-to-Buffer, as in Buffer-to-Buffer credits for FC, a technique to not lose frames.

- BCN: Backward Congestion Notification, a congestion management algorithm.

- Blade Server Chassis: The blade server chassis provides power, cooling, connectivity (passive backplane), and mechanical support for server blades and fabric extenders, which in turn attach to fabric interconnect switch.

- BMC (Baseboard Management Controller).

- Border Ports: Ports on the fabric interconnect that are connected to external LAN, SAN, and management networks.

- California: see UCS and Unified Computing.

- CAM: Computer Array Manager, synonymous of SAM and UCS Manager.

- CEE: Converged Enhanced Ethernet, a term used to indicate an evolution of Ethernet.

- Centrale: Java-based GUI (fat client) for UCS Manager.

- CMC (Chassis Management Controller): This subsystem exists on the fabric extender. It provides a management point for the chassis sensors, fans, power supplies, etc. It also establishes a connection to the Fabric Interconnect and configures fabric extender forwarding elements under its control.

- CNA: Converged Network Adapter, the name of a unified host adapter that support both LAN and SAN traffic.

- CRC: Cyclic Redundancy Check is a function used to verify frame integrity.

- DCB: Data Center Bridging, a set of IEEE standardization activities.

- DCBX: Data Center Bridging eXchange, a configuration protocol.

- DCE: Data Center Ethernet, a term used to indicate an evolution of Ethernet.

- DME (Data Management Engine): The core of UCS Manager, consisting of a transaction engine and an information repository (Management Information Tree).

- DN (Distinguished Name): Immutable property of all MOs that provides a fully qualified (unambiguous name) for the MO.

- dNS: the Fibre Channel domain Name Server.

- DWRR: Deficit Weighted Round Robin, a scheduling algorithm to achieve bandwidth management.

- Enode: a host or a storage array in Fibre Channel parlance.

- F_Port: A Fibre Channel port that connects to Enodes.

- Fabric Extender: a hardware component that connects server blades and management components of the server chassis with fabric interconnect. This subsystem also hosts the CMC.

- Fabric Interconnect: The networking I/O workhorse of the California system. The fabric interconnect is a variant of the Nexus 5000 switch and runs a version of DCOS that is specifically designed for the California application. The UCS Manager runs on top of that DCOS image.

- Fabric Ports: Ports on the fabric interconnect that are connected to the Fabric Extenders.

- FC: Fibre Channel.

- FC_ID: Fibre Channel address, more properly N_Port_ID.

- FC-BB-5: the working group of T11 that is standardizing FCoE.

- FCC: Fibre Channel Congestion Control, a Cisco technique.

- FCF: Fibre Channel Forwarder, a component of an FCoE switch.

- FCIP: Fibre Channel over IP, a standard to carry FC over an IP network.

- FCoE: Fibre Channel over Ethernet.

- FCS: Frame Check Sequence is a function used to verify frame integrity.

- FIP: FCoE Initialization Protocol.

- FLOGI: Fabric Login.

- FPMA: Fabric Provide MAC Address.

- FSPF: Fibre Channel Shortest Path First.

- GUID (Globally Unique Identifier): A generated 16 byte number that has a very high probability of being unique within a certain context.

- HBA: Host Bus Adapter, the adapter that implement the Enode functionality in the host and in the storage array.

- HCA: Host Channel Adapter, the IB adapter in the host.

- HDLC: High-Level Data Link Control, a serial protocol.

- Hypervisor: Software that allows multiple virtual machines to share the same HW platform. See also VM and VMM.

- HOL blocking: Head OF Line blocking is a negative effect that may cause congestion spreading.

- Host Agent: Lightweight agent running on top of a customer-controlled OS (Windows, Linux, ESX), that communicates with SAM. The Host Agent is used for discovery (in particular of interfaces and I/O paths), monitoring of the general health of the server, and optionally provisioning of VLAN, VHBA, and layer-3 interfaces.

- HPC: High Performance Computing.

- IB: Infiniband, a standard network for HPC.

- IBMC (Integrated Baseboard Management Controller): Embedded housekeeping processor in a server used for out-of-band server management. It provides autonomous monitoring, event logging, and recovery control, serves as the gateway for remote system management software, hosts Keyboard-Video-Mouse, and provides remote storage media access. The subsystem provides (host) video controller and super IO functions. IBMCs consist of a processing subsystem that is independent of the computer's main CPU(s), allowing it to be reachable even when the main CPU(s) are powered off or not operational.

- IEEE: Institute of Electrical and Electronics Engineers (www.ieee.org).

- IETF: Internet Engineering Task Force (www.ietf.org).

- IM (Information Model): Domain-specific formal specification of the MO classes (properties, containment, inheritance, ...), rules, and service APIs used by the management framework.

- IMXML (Information Model XML): XML representation of the native information model and API to access and manipulate the IM. This includes mechanisms for getting and setting MOs (RPC), as well as mechanisms to register for events.

- IOC (I/O Consolidation): The ability of a network interface to provide both LAN and SAN connectivity. This effectively indicates that a interface/product/system is capable of Fibre Channel over Ethernet. Aka Unified Fabric.

- IP: Internet Protocol.

- IPC: Inter Process Communication.

- IPMI (Intelligent Platform Management Interface): Defines interfaces and messaging for server out-of-band management. This specification covers transport and underlying protocols that consist of I2C, serial, and RCMP/UDP/IP/Ethernet interfaces. A BMC is an IPMI server.

- ipmitool: Open source project for a utility to manage devices using IPMI 1.5 and IPMI 2.0 (see http://ipmitool.sourceforge.net).

- IPv4: Internet Protocol version 4.

- IPv6: Internet Protocol version 6.

- iSCSI: Internet SCSI, i.e., SCSI over TCP.

- ISO: The International Organization for Standardization is an international standard-setting body composed of representatives from various national standards organizations.

- KVM (Keyboard-Video-Mouse): Industry term that infers access to the keyboard, video, and mouse functions of a system either directly through hardware or remotely as kvm-over-IP.

- LAN: Local Area Network.

- LAPB: Link Access Protocol, Balanced, a serial protocol.

- Layer 2: Layer 2 of the ISO model, also called datalink. In the Data Center the dominant layer 2 is Ethernet.

- Layer 3: Layer 3 of the ISO model, also called internetworking. The dominant Layer 3 is IP, both IPv4 and IPv6.

- Layer 4: Layer 4 of the ISO model, also called transport. The dominant Layer 4 is TCP.

- Layer 7: Layer 7 of the ISO model, also called application. It contains all applications that use the network.

- LLC: Logical Link Control, a key protocol in IEEE 802.1.

- LLC2: LLC used with reliable delivery.

- LLDP: Link Layer Discovery Protocol, an Ethernet configuration protocol, aka IEEE 802.1AB.

- LS (Logical Server): Definition of the identity, PN requirements, connectivity requirements, association policy and storage resources used by a server. A LS is activated on a PN.

- Mezzanine Card: I/O adapter installed on server blades to provide LAN and SAN connectivity.

- MIT (Management Information Tree): Repository of all Managed Object instances, indexed by their Distinguished Names.

- MO (Managed Object): A base class for all objects of the management framework. All MO classes are specified by the IM. All MO instances are stored in the MIT and accessed using their DN or RN.

- MPI: Message Passing Interface, an IPC API.

- N_Port: A Fibre Channel port that connects to switches.

- N_Port_ID: Fibre Channel address, aka FC_ID.

- NFS: Network File System.

- NIC: Network Interface Card.

- NIV (Network Interface Virtualization): Collection of HW and SW (VNTAG and VIC) mechanisms that allow frames to/from vNICs to traverse the same link to a server yet have a consistent set of network policies based on source and target vNICs.

- NX-OS (Nexus Operating System): Embedded Switch Software with L2, L3 and FC protocols, and a Cisco-style CLI. Based on Linux. Note that California doesn't provide L3 forwarding.

- OOB (Out-of-band): Typically applied to a management network deployed in data centers that is separated at some level from the production data traffic. Many customers deploy these are a completely separate physical infrastructure including switches, routers, and WAN feeds.

- PCI: Peripheral Component Interconnect, a standard I/O bus.

- PCIe: PCI Express, the most recent form of PCI.

- PFC: Priority Flow Control, aka PPP.

- PN: Processing Node.

- PNUOS (Processing Node Utility OS): Linux-based pre-boot execution environment that can boot via PXE on a server blade to run diagnostics, report inventory, or configure the firmware state of the PN.

- PPP: Per Priority Pause, aka PFC.

- QCN: Quantized Congestion Notification, a congestion management algorithm.

- RN (Relative Name): Name of an object relative to the name of its container object. Similar to the relative path of a file (where the

DN would be the full pathname of that file).

- R_RDY: Receiver Ready is an ordered set used in Fibre Channel to replenish buffer-to-buffer credits.

- RDMA: Remote Direct Memory Access, an IPC technque.

- RDS: Reliable Datagram Service, an IPC interface used by databases.

- RED: Random Early Detect, an AQM technique.

- RSCN: Registered State Change Notification, an event notification protocol in Fibre Channel.

- SAM: Server Array Manager, synonymous of CAM and UCS Manager.

- SAN: Storage Area Network.

- SCSI: Small Computer System Interface.

- SDP: Socket Direct Protocol, an IPC interface that mimics TCP sockets .

- SFP+: Small Form-Factor Pluggable transceiver.

- SMASH (Systems Management Architecture for Server Hardware): DMTF Standard to manage servers through specific set of CIM Profiles, available through either a Command Line Protocol (CLP), or a web-based transport such as WS-MAN.

- SPMA: Server Provide MAC Address.

- SPT: Spanning Tree Protocol, see also IEEE 802.1Q.

- SR-IOV: Single Root I/O Virtualization, a PCI standard for virtual NICs.

- T11: Technical Committee 11, which is the committee responsible for Fibre Channel (www.t11.org).

- TCP: Transmission Control Protocol, a transport layer protocol in IP.

- TRILL: Transparent Interconnection of Lots of Links Working Group within the IETF.

- Twinax: a twin micro-coaxial copper cable used for 10GE.

- UCS: Unified Coomputing System, aka California.

- UCS Manager: A software providing embedded device management of the entire California system, including the fabric interconnect, fabric extender, and server blade hw elements; the logical servers running within the system; and all of the supporting connectivity information. Synonymous of SAM and CAM.

- Unified Computing: Unified Computing unifies network virtualization, storage virtualization, and server virtualization into one, within open industry standard technologies and with the network as the platform.

- Unified Fabric: see IOC.

- VE_Port: an FCoE port on an FCoE switch used to interconnect another FCoE switch.

- VF_Port: an FCoE port on an FCoE switch used to interconnect an Enode.

- VLAN: Virtual LAN.

- VM (Virtual Machine): Hardware-level abstraction compatible with the underlying server hardware and capable of running a standard operating system (known as the guest OS) on top of a resource scheduler (known as a hypervisor).

- VMM (Virtual Machine Monitor): Hypervisor that allows multiple virtual machines to share the same HW platform.

- VN_Port: an FCoE port on an eNODE used to interconnect to an FCoE switch.

- vHBA (Virtual HBA): Host-visible virtual HBA device (PCI device).

- VIC (Virtual Interface Control): A SW protocol used between a NIC and a switch (in conjunction with VNTAG) to provide Network Interface Virtualization.

- VIF (Virtual Interface). A physical interface of a fabric interconnect supports multiple VIFs that are the policy application points.

- vNIC (Virtual NIC): Host-visible virtual Ethernet device (PCI device).

- VNTAG (Virtual NIC Tag). Layer-2 Ethernet header that uniquely identifies the source vNIC of packets sent by NIC and uniquely identifies the set of destination vNICs of packets sent to the NIC.

- Wireshark: (http://www.wireshark.org/) a public domain protocol analyzer.

- WS-MAN (Web Services for Management): Set of HTTP-based specification to manage devices via CIM Profiles. Includes a specification of transport envelopes (encoding of CIM over XML), event management (WS-Events), security management, ... Implemented in particular by Microsoft MOM as a mechanism for enterprise-class aggregation.

- WWN (World Wide Name): A globally unique identifier assigned to Fibre Channel hosts and ports.

- XAUI: Chip signaling standard for 10 Gigabit Ethernet.

9.2. Figures

9.3. Tables

9.4. Index

Symbols

A

B

C

9.5. Notes

This page is intentionally left blank for notes

This page is intentionally left blank for notes

This page is intentionally left blank for notes